Revising Business Prose

Fourth Edition

Richard A. Lanham
University of California, Los Angeles

New York San Francisco Boston
London Toronto Sydney Tokyo Singapore Madrid
Mexico City Munich Paris Cape Town Hong Kong Montreal

A Pearson Education Company
160 Gould Street
Needham Heights, Massachusetts 02494-2130

Internet: www.abacon.com

Library of Congress Cataloging-in-Publication Data

Lanham, Richard A.
Revising business prose / Richard A. Lanham. -- 4th ed.
p. cm.
Includes index.
ISBN 0-205-30944-5 (paperback)
1. English language--Business English. 2. English language--Rhetoric. 3. Business writing. 4. Editing. I. Title.
PE1479.B87L36 1999
808'.066651--dc21 99-27672
CIP

Printed in the United States of America

10 9 8 7 6 5 4 03 02

CONTENTS

There seems to be something in the water in business schools or at management conferences that destroys people's capacity to speak plainly or write clearly.

John Micklethwait and Adrian Wooldridge, *The Witchdoctors: Making Sense of the Management Gurus*

PREFACE

Revising Business Prose differs from other business writing texts. Let me emphasize these differences up front.

1. About *revising*. As its title indicates, it is about *revising*; it does not deal with original composition. Students may agonize about *what to say* but in the workplace this "student's dilemma" hardly exists. The facts are there, the needs press hard, the arguments lie ready to hand, the deadline impends. The first draft assembles itself from the external pressures. Then the sweat really begins—*revision*, commonly done in group settings. Collective revision up through a hierarchy determines the final text for much business writing. It offers the choice leverage for improvement but it needs detailed shared guidance. That detailed guidance deserves a special book. "All writing is rewriting," goes the cliché. OK. Here's the book for it. It offers a collective writing philosophy which a group can easily and quickly learn to share.

2. Translates "the Official Style" into English. A specific analytical and social premise informs *Revising Business Prose*. Much bad writing today comes not from the conventional sources of verbal dereliction—sloth, original sin, or native absence of mind—but from stylistic imitation. It is learned, an act of stylistic piety which imitates a single style, the bureaucratic style I have called "the Official Style." This bureaucratic style dominates written discourse in our time, and beginning or harried or fearful writers adopt it as a protective coloration.

So common a writing pattern deserves a separate focus, a book of its own.

3. **Rule-based and self-teaching.** Because it addresses a single discrete style, *Revising Business Prose* can be *rule-based* to a degree which prose analysis rarely permits. This set of rules—the *Paramedic Method* (PM)—in turn allows the book to be self-teaching. In the workplace, the pressures animating revision rarely permit time off to take a special writing course. You need something useful right then. Again, a special book for this purpose seems justified.

4. **Useful in all jobs but aimed at business.** Because the Official Style flourishes in university, workplace, and government, the book can work in all these contexts. The Official Style afflicts us all, a common plague. The fourth edition of *Revising Business Prose*, though, is aimed at the Official Style in business and chooses its examples accordingly.

5. **Perfect for government-mandated "plain English" revisions.** The government and the courts have begun to require that documents be written in plain English rather than in legal terminology or its kissin' cousin, the Official Style. The Securities and Exchange Commission has issued a *Plain English Handbook* on how to do this. And, most recently (1 June 1998), the President of the United States signed an Executive Memorandum mandating that documents written for the American people should be written in plain English. It is ironical enough that the government, the ever-fresh source of the Official Style, has taken a stand against it. No matter. If you must comply with such "plain language" requirements, *Revising Business Prose* is the perfect book for you. Since it first appeared in 1981, it has been teaching people to do precisely this kind of revision.

6. **Based on the electronic word.** If writing on an electronic screen has revolutionized prose composition and prose style, nowhere has the revolution hit harder than in revision. Revision is much easier to do on screen and most of it nowadays is done there. Electronic text brings with it a new stylistic theory as well as new means of moving words around on an expressive

surface. I have tried to infuse the fourth edition of *Revising Business Prose* with an awareness of this new writing medium.

7. Saves time and money when used as directed. The bottom line, short and long term: the Paramedic Method, when used as directed, saves time and money. Lots of it. The Lard Factor of the Official Style usually runs about 50%. Add up everything the written word costs you for a year, time and materials, and divide by two. That figure will guesstimate pretty well the savings *Revising Business Prose* can generate.

Cutting communications costs in half generates gigantic savings. An attorney for the Environmental Defense Fund has suggested that the federal courts allow documents to be filed when printed on both sides of the paper. This would save 75 million pieces of paper a year. That is 50% of the 150 million pieces filed in the federal courts each year. *Revising Business Prose*, at its best, offers a comparable 50% saving. In today's fierce cost-cutting environment, an employee who can generate savings of this magnitude in her own work, and in revising the work of others, adds immediate value to the enterprise, and in a way which shows.

But the book has an even larger end in view—stylistic self-consciousness. This verbal self-awareness, however generated, is like riding a bicycle. Once learned, never forgotten. And stylistic self-consciousness changes how we read and write not only in a single bureaucratic register but across the board. From a particular focus, this book aims to encourage a general self-consciousness about words.

8. Sentence-based. *Revising Business Prose* focuses on the single sentence. Get the basic architectures of the English sentence straight, I think, and everything else follows much more easily. If you can understand the basic architecture of the sentence, you can generalize upward to the structure of a paragraph, a letter, a memo. We're analyzing in this book the microeconomics of prose. All our work together will be close-focus writing and what acousticians call "near-field" listening. Such close-focus work is as seldom performed as it is universally needed. We spend a great deal of time worrying about our verbal "P's and Q's" which we ought to spend worrying about our sentence architecture. That's where the big misunderstandings occur. Getting the individual sentence straight is *the best way to begin.*

9. Provides Emergency Room treatment. I've called my basic procedure for revision the Paramedic Method because it provides emergency therapy, a first-aid kit, a quick, self-teaching method of revision for people who want to translate the Official Style, their own or someone else's, into plain English. But it is only that—a first-aid kit. It's not the art of medicine. As with paramedicine in underdeveloped countries, it does not attempt to teach a full body of knowledge but only to diagnose and cure the epidemic disease. It won't answer, though at the end it addresses, the big question: having the cure, how do you know when, or if, you should take it? For this you need the full art of prose medicine, a mature and reflective training in verbal self-awareness. I've addressed this larger stylistic domain in another book, *Analyzing Prose.*

People often argue that writing cannot be taught, and if they mean that inspiration cannot be commanded nor laziness banished, then of course they are right. But stylistic analysis—revision—is something else, a method, not a mystical rite. How we compose—pray for the muse, marshal our thoughts, find willpower to glue backside to chair—these may be idiosyncratic, but revision belongs to the public domain. Anyone can learn it.

And follow the PM. It works only if you *follow* it rather than *argue* with it. When it tells you to get rid of the prepositional phrases, get rid of them. Don't go into a "but, well, in this case, given my style and my company, really I need to ..." bob and weave. You'll never learn anything that way. The PM constitutes the center of this book. Use it. It is printed on a separate page in the front. Clip it out and tack it above your desk for easy reference.

10. What's new with the fourth edition?

- New organization
- New examples
- A set of interactive exercises, with discussion of each exercise and suggested revisions, is now available from Rhetorica, Inc., 927 Bluegrass Lane, Los Angeles, California 90049 (tel. 310-472-1577; fax 310-472-4757; email: rhetorica@aol.com).
- The *Revising Business Prose* video, a 40-minute video using animated print, sound, and color to depict revision in action, is once again available, also from Rhetorica, Inc.

A word on the last two items. Prose revision is an interactive process *par excellence,* best demonstrated and practiced dynamically. But the print medium—this book—can only describe it one step at a time. The interactive exercises allow you to try out, view on screen, various possibilities as you revise. The video shows revision as it happens.

In this book, as in all my work, the editorial and scholarly eye of my wife, Carol Dana Lanham, has spared both reader and author many inconsistencies, gaffes, and stupidities. *Gratias ago.*

R. A. L.

The Paramedic Method

1. Circle the prepositions.
2. Circle the "is" forms.
3. Ask, "Where's the action?" "Who's kicking who?"
4. Put this central action in a simple active verb.
5. Start fast—no slow windups.
6. Write out each sentence on a blank screen or sheet of paper and mark off its basic rhythmic units with a "/".
7. Mark off sentence lengths in the passage with a big "/" between sentences.
8. Read the passage aloud with emphasis and feeling.

CHAPTER 1

WHO'S KICKING WHO?

What should business writing be like? It ought to be fast, concrete, and responsible. It should show *someone acting*, doing something to or for someone else. Business life offers few occasions for the descriptive set-piece; it chronicles history in the making, depicts someone working on matter or with people. It seldom relates abstract concepts for the fun of it; abstractions occur as parts of a problem to be solved. Business prose ought, therefore, to be *verb-dominated* prose, lining up actor, action, and object in a causal chain and lining them up fast.

Often, though, business prose moves in the opposite direction, toward a special language we might call "the Official Style." The Official Style is the language of bureaucracies, of large organizations; it is a noun-centered language, full of static abstractions, voiced always in the passive, and slow. Above all, it strives to *disguise the actor*, allow such action as cannot be quashed entirely to seep out in an impersonal construction—never "I decided" but always "It has been decided that...."

It isn't hard to see why the Official Style threatens to conquer the business world as it has done the world of government. We are all bureaucrats these days, or shortly will be, whether we work for the government directly or soldier on in the private sector and get our government money through grants, contracts, or subsidies. And even if—especially if—we belong to that shrinking part of the private sector that remains truly private, we'll be for certain filling out government forms, having OSHA for lunch whether we invited her or not.

In spite of a growing chorus of dissent, to which *Revising Business Prose* has lent a voice since 1981, the Official Style continues to thrive. It has colonized a new land—management theory. As a recent study of management theorists remarks, "There seems to be something in the water in business schools or at management conferences that destroys people's capacity to speak plainly or write clearly."[1] The "something" is the Official Style, the new learned language of professional bosses.

Not all business communication employs the Official Style, of course. Good prose is to be found across the spectrum, from interoffice memos to articles in the *Harvard Business Review*. Nevertheless, it sometimes seems as if business writing nowadays is going in one direction and management practice in another. As management becomes more horizontal, less bureaucratic, the language MBAs and business gurus use to describe this management becomes more bureaucratic and hierarchical. Management, now that it sees itself as a learned profession, has begun to ape the special languages of the other learned professions. Employees exposed to repeated doses of this special language have, in self-defense, invented a game called "Buzzword Bingo" which plays upon the buzzwords used by the speaker.

---

[1]*The Witchdoctors: Making Sense of the Management Gurus*, John Micklethwait and Adrian Wooldridge (New York: Times Books/Random House, 1996), p.12.

You'd think that business writing would be moving in the opposite direction. After all, we are living in an "information age." Communication of all sorts, but especially within the individual firm, has never been more valued. Nor has the need to democratize information, and the language which carries it, to ensure that it reaches employees newly "empowered" to decide how to do their own jobs. Yet the tide of what, in this book, I call the Official Style rises ever higher. For every drop of plain-language law, there is a five-gallon bucket of MBA-speak and guru psychobabble. What's going on? I'm not sure, but the paradox needs to be explained. What is clear—clearer even than with earlier editions of this book—is the need to translate the Official Style into plain English.

We all have to do business in the Official Style—Federalese, Bureaucratese, Sociologese, Educationese, Doublespeak, or just our own firm's "company style." And to do business in it, we will often—though not always—want to translate it into English. If "initiation of the termination process is now considered appropriate" *re* us, if we have been "downsized," "outplaced," "de-employed," or "proactively outplaced," we must know that it's time to look for another job. And some of us may also practice such translation in the name of business efficiency, verbal aesthetics, or plain cultural sanity.

And, as a final indignity, we toil in a society so fond of lawsuits that we want to write everything in legal language—or at least in language that *sounds* legalistic—to cover our own backsides in the likely event that, somewhere down the line, we'll get sued. But, in a double-bind strategy that would do credit to the IRS, the government also nowadays passes plain-language laws which prohibit writing the defensive legalistic double-talk which the government itself retails by the cubic ton.

In trying to escape from this maze, we'll begin with some nuts-and-bolts details of sentences: actor, action, shape, rhythm. These details lead to a discussion of voice and authority in business prose. After that, we'll consider the Official Style as a whole: What is it? What does it do? How has it come about? How does it fit into an information economy in which the scarce commodity is not information but the human attention needed to make sense and use of it?

WHERE'S THE ACTION?

Suppose you want to describe a simple action: "Jim kicks Bill." That would be too direct, unfeeling, unceremonial, for an Official Stylist—actor, action, object, stand out clearly. The Official Style prompts us instead to write: "One can easily see that a kicking situation is ongoing between Jim and Bill." Or, "This is the kind of situation in which Jim is a kicker and Bill is a kickee." Jim cannot enjoy giving Bill a good boot; no, for Official Use, it must be "Jim is the kind of individual who enjoys participation in an interactive kicking situation *re* Bill." Actor, action, object, all blur. Absurdly contrived example? Here's a real one that reveals the root effort of the Official Style—to convert action into stasis.

> The bond markets are in disbelief of the ability of First world countries to maintain this level of debt.

The *actor* stands clear enough—"The bond markets." But what are the bond markets doing? They are *not believing*. But that action—not believing—is blurred by being made into a noun—"disbelief"—coupled with the verb "to be," giving us "are in disbelief." The noun "disbelief" then requires two prepositional phrases—"*of* the abilities *of* First world countries." These prepositional phrases require, in turn, that it takes much longer to reach *what the bond markets don't believe*—that the First world can maintain a debt level. Where's the action?

> The bond markets doubt that the First world countries can maintain this debt level.

Instead of

> are in disbelief
> *of* the ability
> *of*

we say

> doubt.

One word instead of six, positive instead of negative (*doubting* instead of *not believing*, zero prepositional phrases instead of two. Most important, we've found the *action*.

Where is it in this observation by a computer pioneer?

> The lack of usability of software and poor design of programs is the secret shame of our industry.

The action lies in *shame*. Get rid of "is the secret shame of" and use "shame" as the verb. OK. Where's the *actor*? Two of them: "The lack of usability of software" and "poor design of programs." Look at each in turn.

> The lack
> *of* usability
> *of* software

Again, the prepositional phrases blur the central assertion: something is being said about *using software*. What? It is *unusable*. The Paramedic Method tells you to get rid of the prepositional phrases. When we do so, we have a natural subject: *unusable software*. Once we see this, revising the second subject becomes easy. Instead of "poor design of programs" we have "poorly designed programs." And so the revision:

> *Original*
>
> The lack of usability of software and poor design of programs is the secret shame of our industry.

> *Revision*
>
> Unusable software and poorly designed programs shame our industry.

The villains are "is" and those prepositional phrases. Get rid of them and you see who is doing what to whom. Here's a perfect example:

> An extension of the concept of wireless access to commercial information services is the provision of the same type of access to corporate databases.

See how it breaks out?

> An extension
> *of* the concept
> *of* wireless access
> *to* commercial information services
> *is* the provision
> *of* the same type
> *of* access
> *to* corporate databases.

Behold the basic Official Style formula: "is" surrounded by strings of prepositional phrases. Once you learn this formula, you can write it in your sleep and people often, as here, do just that. A vague noun plus blah-blah-blah is a vague noun plus blah-blah-blah. "An extension…is the provision…"—what does this mean? You have to follow out all those prepositional phrases to see Who is kicking Who. The root *action* seems to lie buried in "extension." Let's start from there.

> Something *extends* something.

OK. What can be extended? "Wireless access."

> Wireless access can be extended from commercial databases to corporate ones.

It still doesn't make a lot of sense. The writer seems to have meant "Wireless access to commercial databases suggests access to corporate ones as well." If so, both *actor* and *action* are so deeply buried it requires a root canal to find them. This sentence shows thought not only imprisoned by the Official Style but stultified by it.

Here's a yet longer sentence which blurs *actor, action,* and *object*.

> In light of the pervasive problem of overcrowding at UC Lone Pine, providing another coffee house on campus would offer the university's growing population some kind of compensatory convenience.

Consider, first, all the *possible* actions this sentence invokes:

pervades
overcrowds
provides
offers
grows
compensates

The reader can't decide which should predominate. And who is *acting* in this mixture of possible actions? The "pervasive problem"? "UC Lone Pine"? "Another coffee house"? To decipher a sentence like this, we have to take it part by part. Let's start with the opener:

In light
of the pervasive problem
of overcrowding
at UC Lone Pine

Notice how this string of prepositional phrases delays the real action of the sentence? The Paramedic Method uses prepositional phrases as a way into a sentence, a place to begin. Can we get rid of them here? Let's try a preemptive strike:

"Overcrowded Lone Pine." Three words instead of ten. And a natural *subject*—UC Lone Pine. OK. Now, what *action*, what *verb*, fits UC Lone Pine? Now it is easy to see: Lone Pine *needs* something. Now that actor and action are clear, the object is clear, too—*another coffee house.*

Original

~~In light of the pervasive problem of~~ overcrowd[**ed**]~~ing at~~ UC Lone Pine~~, providing~~ **needs** another coffee house. ~~on campus would offer the university's growing population some kind of compensatory convenience.~~ (29 words)

Revision

Overcrowded UC Lone Pine needs another coffee house. (8 words)

Eight words instead of 29. A Lard Factor of 72%.

We shall use the Lard Factor as a rough numerical approximation of success in revision. You find it by dividing the difference between the number of words in the original and the revision by the number of words in the original. In this case:

29 − 8 = 21 ÷ 29 = 72%

We've found *actor, action, object.* And the rest of the original? Pure lard, generated by a writer trapped in the Official Style. The Official Style, by hiding the sentence's natural subject and verb in a swamp of nominal constructions and prepositional phrases, allows us to play *Let's Pretend.* It's not a good game for business but it is played a lot. To escape it, ask "Where's the *action*? Somebody's doing something to someone else. Who? *Who's kicking who*?" (I know it should be *whom* but doesn't it sound stilted?)

All right. The Official Style displaces the action from verb to noun, aims to convert *action* into *stasis.* In doing so, as we have seen, it employs two debilitating stylistic patterns, the prepositional-phrase string and the slow sentence windup. Let's look at these two patterns in more detail, starting with the slow windup. I'll christen it the "Blah blah *is that*" opening.

THE "BLAH BLAH *IS THAT*" SENTENCE OPENING

Some specimens from my collection:

What I would like to signal here *is that*...

My contention *is that*...

What I want to make clear *is that*...

What has surprised me the most *is that*...

The upshot of what Jones says here *is that*...

The point I wish to make *is that*...

What I have argued here *is that*...

My opinion *is that* on this point we have only two options...

My point *is that* the question of...

The fact of the matter *is that*...

My own personal view *is that*...

It *is* important to keep in mind *that*...

It's a snap to fix this pattern: amputate it. Eliminate the mindless preludial fanfare. Start the sentence with whatever follows "Blah blah *is that*...." On a word-processor it couldn't be simpler: do a global search for the phrase "is that" and revise out the offending phrase each time.

By amputating the fanfare, you *start fast*. And a fast start may lead to major motion. That's what we're after. Finding the *action*.

> *Throughout the ABC organization, however, there is a strong belief that* decisions should be made by those closest to the problem. (21 words)
>
> ABC believes that decisions should be made by those closest to the problem. (13 words; Lard Factor 38%)

"There is a strong belief that" = "believes" and all follows from that.

> *Despite the fact that* many truckers feel the railroads cannot cut substantially into freight hauls that have belonged to trucks for years, the rails are doing it. (27 words)
>
> Although many truckers think railroads cannot cut into established truck hauls, they are doing it. (15 words; LF 44%)

The real action here lies in *they are doing it*; the quicker we get there, the better. Another easy amputation:

> *Since it is the case that* construction is soon to be finished on the $50 million Lovenest Hotel building... (19 words)
>
> Since the $50 million Lovenest Hotel is nearly finished... (9 words; LF 53%)

Prepositional-Phrase Strings + "Is"

The prepositional-phrase strings usually pose a more complex challenge than the "Blah blah *is that*" opening, but still they are easy to fix.

> Hypertext was invented to facilitate the process *of* navigating *through* a presentation *of* a collection *of* interrelated topics.

Diagrammatically:

> Hypertext was invented to facilitate the process
> *of* navigating
> *through* a presentation
> *of* a collection
> *of* interrelated topics.

Again, notice the plethora of possible actions:

> *facilitate*
> *navigate*
> *present*
> *collect*
> *interrelate*

Surely *navigate* supplies the main action. The Official Style loves the words "facilitate" and "process." You can almost always get rid of them:

> Hypertext was invented to ~~facilitate the process of~~ navigate...

Now for

> *through* a presentation
> *of* a collection
> *of* interrelated topics.

This string of prepositional phrases says the same thing three times: a presentation of a collection of interrelated topics. If you insist on breaking up that prepositional-phrase string, you soon see that all you need is "interrelated topics."

> Hypertext was invented to facilitate the process of navigating through a presentation of a collection of interrelated topics. (18 words)

> Hypertext was invented to navigate through interrelated topics. (8 words; Lard Factor 55%)

Notice how boring and monotonous that string of prepositional phrases becomes? The sentence wants to describe a causal process, as so much of business writing does. What *caused* A? How get from A to B to C? The Official Style always does its best to obscure a causal pattern or a chain of command. In this example, the chain runs from "hypertext" to "navigate" to "interrelated topics." "Navigate" is the crucial action. Action verbs should stand at the center of business writing. You should always know who is acting, how, when, and toward whom.

In the next sentence, the writer can't bear to make clear any of these relationships.

> Dear Brad: We have had an acceptance of our offer of employment from the outstanding candidate for the position of Regional Marketing Manager.

The string of prepositional phrases tips you off that something has gone badly wrong:

> We have had an acceptance
> **of** our offer
> **of** employment
> **from** the outstanding candidate
> **for** the position
> **of** Regional Marketing Manager.

Where's the *action* here? The actor is clear—"We"—but what is the central action? What have "We" done? They have **hired** a new Regional Marketing Manager and it isn't Brad. But

"We" can't bring him or herself to say so—"We have hired someone else as Regional Marketing Manager" (9 words instead of 21; LF 54%). So this central action is blurred through the strings of prepositional phrases. But you don't make bad news better by stringing it out like saltwater taffy.

The writer continues to writhe within the confines of the Official Style's prepositional-phrase strings in the next sentence.

> Nevertheless, I extend
> **to** you my very best wishes
> **for** a successful conclusion
> **to** your personal search
> **for** the right career path
> **for** you.

A "personal" search? What other kind is there? The right career path "for you." For who else? When I "extend my very best wishes," I "wish." The writer desperately wants to avoid saying the plain truth: "Nevertheless, I wish you success in your job search" (LF 62%). The Official Style provides the perfect vehicle for avoiding the plain truth—a truth which would be much kinder than the Official Style flummery. It is those prepositional phrases which—in a sentence of the sort of which this is a type of—reveal the real problem.

THE PARAMEDIC METHOD

These examples have given us the beginnings of the Paramedic Method (PM):

1. Circle the prepositions.
2. Circle the "is" forms.
3. Ask, "Where's the action?" "Who's kicking who?"
4. Put this *action* into a simple active verb.

Now, let's take this search for actor and action, for a detectable and responsible *chain of command* in a sentence, one step further.

> After reviewing the research and in light of the relevant information found within the context of the conclusions, we feel that there is definite need for some additional research to more specifically pinpoint our advertising and marketing strategies.

The standard formula: "is" + prepositional phrases fore and aft. And often a "to" infinitive sign joins the conga line.

After reviewing the research and
in light
of the relevant information found
within the context
of the conclusions,
    we feel that there *is* definite need
for some additional research
to more specifically pinpoint our advertising and marketing
        strategies.

Who's kicking who? Well, the kicker is obviously "we." And the action? *Needing*—but buried in *there is definite need.* So the core of the sentence emerges as "We need more research." Let's revise what comes before and after this central statement.

of previous
~~After reviewing the~~ research ~~and in light of the relevant infor-~~
suggest
~~mation found within the context of~~ the conclusions~~, we feel that~~
that we **more**
~~there is definite~~ need ~~for some additional~~ research to ~~more spec-~~
~~ifically~~ pinpoint our advertising and marketing strategies.

The revision then reads:

> The conclusions of previous research suggest that we need more research to pinpoint our advertising and marketing strategies.

Eighteen words instead of 38—LF 53%. Not bad—but wait a minute. How about "the conclusions of"? Do we need it? Why not:

> Previous research suggests that we need more research to pinpoint our advertising and marketing strategies. (LF 60%)

And this revision, as so often happens, suggests a further and more daring one:

> **has failed**
> Previous research ~~suggests that we need more research~~ to pinpoint our advertising and marketing strategies. (LF 71%)

By now, of course, we've changed kicker and kickee and, to an extent, the meaning. But isn't the new meaning what the writer wanted to say in the first place? A previous failure has generated a subsequent need? And the new version *sounds* better, too. The awkward repetition of "research" has been avoided and we've finally found the real first kicker, "previous research," and found out *what it was doing*—it "failed." We can now bring in the second kicker in an emphatic second sentence:

> Previous research has failed to pinpoint our advertising and marketing strategies. *We need to know more.*

No "is," no prepositional phrases, a LF of 58%, and the two actors and actions clearly sorted out. The drill for this problem stands clear. Circle every form of "to be" (e.g., "is," "was," "will be," "seems to be") and every prepositional phrase. Then find out who's kicking who and start rebuilding the sentence with that action. Two prepositional phrases in a row turn on the warning light, three make a problem, and four invite disaster. (As you read on, see if you catch me disregarding my own advice, and revise accordingly.) With a little practice, sentences like:

> The role of markets is easily observed and understood when dealing with a simple commodity such as potatoes.

will turn into:

> A simple commodity like potatoes shows clearly how markets work. (LF 44%)

Action Programs and Wasting Money

Every sentence describes an action program. To stay in business, you must not lose sight of *what's happening*, who is doing what to whom. Here is a short sentence that manages to blur every important action:

> The trend in the industry is toward self-manufacture by some companies of their own cans, and packing technology is changing packaging requirements so as not to require the typical heavy metal can.

First, notice that the sentence falls into two unrelated parts:

> The trend in the industry is toward self-manufacture by some companies of their own cans,
>
> and
>
> packing technology is changing packaging requirements so as not to require the typical heavy metal can.

The two parts must bear some relation to each other; if not, why do they share the same sentence? What might it be? In cases like this, best to find the action in each half and then guess at the relationship between them.

So, the first half. Where's the *action*? Hidden down there in *self-manufacture*. Somebody is *making* something. Who is doing it? Who is the subject? We have a choice of three: *trend, industry, companies.* Easy to see the natural subject here: *companies.* We now have a clear action program:

> Some companies make their own cans.

We know where the action is, who is kicking who. The dead-rocket opening, *The trend in the industry is toward* = *now*, and so we have:

> Some companies now make their own cans. (LF 53%)

The LF of 53% only begins to tell the story. The writer could not express a basic simple action, could not see who was acting or what the action was. And the stumbling approximation took twice as long.

Am I over-emphasizing a trivial mistake? Suppose your company could save 50% of everything you spent on the written word, from paper and ink to fax machines and employee time? And what of the mental processes of someone who cannot see, or explain, a simple business process? The idea of a *company* should pose no special intellectual challenge to a business person. Nor should *making*, nor even *cans*. We're not talking about quantum theory. The gigantic wastage revealed by such a sentence (or half a sentence—we are far from finished with it) comes in the *thinking* revealed. Thinking time costs a lot of money.

Now for the second half of the sentence:

> packing technology is changing packaging requirements so as not to require the typical heavy metal can.

Again, who's kicking who? Where is the *action*? The writer tips us off by his redundancy: *requirements* and *require.* OK, we have a verb—*require.* Who or what is doing the requiring? Not, as the writer writes it, the *requirements*, since that means that the *requirements* are *requiring*, which is tautological and witless. No, our actor is *technology*. Now, at last, we have actor and action: *packing technology requires*. So we come to this:

> packing technology no longer requires a heavy metal can. (LF 44%)

At this point, we can contemplate the relationship between the two halves of the sentence. Since the writer did not specify a relationship but merely juxtaposed the two parts, we must guess. My guess:

> Since packing technology no longer requires a heavy metal can, some companies now make their own.

Our Lard Factor, as chance would have it, scores a bull's-eye: 50%, 16 words instead of 32. But, again, the LF only uncurtains the problem. Prose models thought, and thought must account for events and the causal relationships between them, the causes and results that keep a business solvent. The Official Style attacks business at its heart—a knowledge of *what's happening*.

THE PM *SUPER-SLIM WEIGHT LOSS PROGRAM*

Now for an example from a current high fashion—the consultant's report. This consulting company was developing a "strategic market plan" for a bank.

> The purpose of an environmental scan is to obtain a general understanding of the external business environment we are currently in and expect to be in over the near-term. This may include any number of factors, but they are factors that may significantly impact

> the bank's business, either positively or negatively depending upon how we manage our way through them.

OK. Not hard to understand, but blurred and, my goodness, *wordy*. Let's do one of our PM diagrams to get inside the problem.

> The purpose
> *of* an environmental scan
> **is**
> *to* obtain a general understanding
> *of* the external business environment
> we are currently *in*
> and expect to be *in*
> *over* the near-term.
> **This** may include any number
> *of* **factors,**
> but they are **factors** that may significantly impact the bank's business,
>   either positively or negatively
>   depending upon how we manage our way through them.

I've taken two liberties to emphasize the Official Style formula: (1) I've included the infinitive sign "to" as if it were a preposition, since, here and usually, it acts like one in the Official Style; (2) I've used a hollow, outline typeface for "This" since we don't know what it refers to, "factors" since they compose the "this." This blurred reference blurs the relationship between the two sentences.

To work with the PM: identifying the prepositions and the "is" establishes the Official Style formula:

1. Slow windup, with a prepositional phrase or two, ending with **is**:

> "The purpose of an environmental scan **is**..."

2. A string of prepositions/infinitives following:

> "to obtain...currently *in*...expect to be *in*...*over*...the near"

"This initial phase of our analytical methodology has facilitated our vision that"—as an Official Stylist might put it, or, as we might say in ordinary English, "We can now see that"—*actor* and *action* have gone into hiding, as usual. The grammatical subject of the sentence is "purpose," but is "purpose" the central actor? No, it is *scan*. The passage wants to tell us what an "environmental scan" *does*. Where is the *action*? In hiding, too, in "to obtain a general understanding of...." "To obtain a general understanding of" = "to understand." Now we have a central action—*understand*. But wait a minute. Who is obtaining the understanding? As written, it is the *scan*, but that makes no sense. The scan doesn't do the understanding, we do, using the scan. The writer has confused cause—*scan*—and effect—*understanding*. Applying the PM's first three rules diligently has allowed us to see what the problems are, where the sentence goes wrong.

OK. We do have an *actor*—an environmental scan. What is it *doing*? Well, it can't "obtain a general understanding," since only people can understand, not reports. What word might we use for "obtaining a general understanding of"? My favorite thesaurus[2] suggests: "observe, scrutinize, regard, contemplate, consider, review, study, examine, investigate, pore over, peruse, appraise, assess, size, up, survey,..." and many more. I'll go for "survey."

OK. We have—and it hasn't been easy—an *actor* and an *action*:

> An environmental scan surveys...

What does it survey? As written:

> the external business environment we are currently in and expect to be in over the near-term.

[2]J. I. Rodale, *The Synonym Finder* (Warner Books, 1978).

As always, take it a part at a time:

> the external business environment we are currently in
> = the current business environment
>
> and expect to be in over the near-term = foreseeable
> = the current and foreseeable business environment.

Now we have a before-and-after photo:

> *Original*
>
> The purpose of an environmental scan is to obtain a general understanding of the external business environment we are currently in and expect to be in over the near-term. (29 words)
>
> *Revision*
>
> An environmental scan surveys the current and foreseeable business environment. (10 words; LF 65%)

Our Lard Factor does better than our bull's-eye one half—65%. But more important, the revision has identified the real actor and the real action, and clarified the relationship between them.

A Pause for Reflection

Before we take on the second sentence from this consultant's report, let's reflect upon the first. Mistaking actor and action and the relationship between the two does not constitute a trivial error, a "matter of style." Business may not be war without the guns, as some say it is, but it does involve players and struggles between them. You should know who is doing what to whom. To confuse the players and the game constitutes a *big* mistake. And who is making it? A *consultant.* Somebody who is *advising someone else how to act.* And this advisor doesn't know who is doing what to whom.

ONWARD

Time for the consultant's second sentence.

> This may include any number of factors, but they are factors that may significantly impact the bank's business, either positively or negatively depending upon how we manage our way through them.

The *actor*? "Factors," obviously. But "factor" is a *very* general word (and a favorite in the Official Style—that's why I chose it for "Lard Factor"). What *exactly* does it mean here? To find out that, we must know what "This" is. "This," whatever it is, includes all those "factors" and they are vital to the sentence. So "This" is vital. But alas it is a **This** this. It refers back to something in the previous sentence but we don't know what. According to the rules, it refers back to the noun most immediately preceding, which would be "near-term." But that doesn't make much sense. So what does it refer to? I'll indicate the "possible perps" as the police like to say ("perp" = perpetrator).

> The **purpose** of an **environmental scan** is to obtain **a general understanding** of **the external business environment** we are currently in and expect to be in over **the near-term.**

Does **This** refer to **a general understanding** or to **the external business environment**? Your guess is as good as mine. The reference oscillates between the one and the other. I choose the "business environment." Elements of the business environment affect ("impact" is the Official Style word for "affect" or "influence") the bank's business. Somehow. So the sentence should read:

> [The business environment] may include any number of factors, but they are factors that may significantly impact the bank's business, either positively or negatively depending upon how we manage our way through them.

Wait a minute. Things get worse and worse. There may be any number of factors (whatever they are) and they *may* affect, or may not, the bank's business, either up or down. Depending on what the bank does. At this point in the analysis, your mind begins to lose its bearings. The sentence, which aims to guide us, has lost us instead. There are a bunch of factors, or may be, in the business environment and they may affect the bank, or may not, depending on what the bank does. So what else is new? Under the infarcted language lurks a trite commonplace:

> How these factors [or "this environment"] affect the bank's business depends on how the bank responds to them.

Now let's put the two sentences together:

> An environmental scan surveys the current and foreseeable business environment. How this environment affects the bank's business depends on how the bank responds to it. (25 words)

The original again:

> The purpose of an environmental scan is to obtain a general understanding of the external business environment we are currently in and expect to be in over the near-term. This may include any number of factors, but they are factors that may significantly impact the bank's business, either positively or negatively depending upon how we manage our way through them. (60 words)

A Lard Factor of 58%. But that fraction only points toward the real trouble. The two sentences don't say much. The Official Style fools everybody connected with it—writer as well as reader—into thinking more has been thought and said, and paid for, than in fact has been. We're not talking about fancy phrases to make us feel important—"Return the handset to the cradle" instead of "hang up," or "achieve the desired velocity" instead of "go as fast as you want." We are talking about self-delusion.

Now for another kind of consultant's report. In recent years, ordinary business writing has become so infected by the Official Style that a new scholarly specialization has emerged to teach business writing. Teachers working in this field, like other academics, must publish articles, and these articles must be acceptable to two groups devoted to the Official Style—the MBA types and the empirical linguists. Not only how business is done but how it is studied now goes forward in the Official Style. The following example, which is *about business writing*, hides actor and action under a slime of passives, prepositions, verbs made into nouns, and Latinate "shun" words—institu*tion*alized, organiza*tions*, configura*tions*. The resulting prose looks like this:

> With respect to institutionalized properties of organizations, our framework suggests that over time, the actions exercised by humans in the domains of media use, message structure, and language become habitual, and particular configurations of media, message structure, and language emerge and are invoked in certain circumstances to achieve some communicative intent.

Action? Action? Actor? Actor? What are they talking about? Let's call up the PM to find out:

1. Circle the prepositions.
2. Circle the "is" forms.
3. Ask, "Where's the action?" "Who's kicking who?"
4. Put this "kicking" action in a simple (not compound) active verb.
5. Start fast—no slow windups.

OK, we invoke rules 1 and 2:

> *With* respect *to* institutionalized properties *of* organizations, our framework suggests that *over* time, the actions exercised *by* humans *in* the domains *of* media use, message structure, and language **become** habitual, and particular configurations *of* media, message

> structure, and language emerge and **are** invoked *in* certain circumstances *to* achieve some communicative intent. (51 words)

Now, Rule 3. Who is kicking who? Where's the *action*? It's almost impossible to see, but the sentence does contain two centers of power and we can start there:

> actions become habitual
>
> configurations of media emerge and are invoked

In these two centers of power lurk five possible actions:

> *act*
> *habituate*
> *configure*
> *emerge*
> *invoke*

Let's translate the individual phrases from the Official Style into English:

> *institutional properties of organizations* = **company habits**
>
> *actions exercised by humans in the domains of media use, message structure and language become habitual*
> = **people tend to express themselves in habitual ways**
>
> particular configurations of media, message structure, and language emerge and are invoked in certain circumstances to achieve some communicative intent
> = **people tend to express themselves in habitual ways**

Have I done the passage an injustice? I don't see how. The plain English for the whole passage, so far as I can see, must run something like:

> In large organizations, people tend to express themselves in habitual ways. (11 words; LF 78%)

You might then specify what those ways, those media, are. But you would have to name names, have specific actors performing specific actions, people writing, speaking, sending E-mail, following accepted report forms, whatever.

Anybody who has studied the social sciences has read acres of such prose. It may be all right for that kind of thinking—though I do not think so—but for business? As a model for business prose, I find a passage like this *scary*. It muddles the world of affairs, a place where *actions* and their *actors* must be clear, and the causal relationships between them precise.

A Pause for Reflection

In this first chapter of *Revising Business Prose*, we have considered three kinds of writing: business writing, writing about business, and writing about business writing. They all exhibit the same formulas—the formulas of the Official Style. To examine them, we have put together the first half of the Paramedic Method:

1. Circle the prepositions.
2. Circle the "is" forms.
3. Ask, "Where's the action?" "Who's kicking who?"
4. Put this "kicking" action in a simple active verb.
5. Start fast—no slow windups.

The Paramedic Method takes time. Yes, of course it does. PM revision takes hard work too. But that hard necessity only reveals the magnitude of the problem. We are talking about big mistakes, not fine details. A recent police sting up in Silicon Valley recovered a hoard of stolen chips. The police officer who spoke to the press wanted to point out that what they had seen was only the tip of the iceberg, the cream on the coffee, the surface of a big pond. So he said, "What you see here? It amounts to comparably nothing." So with the examples we have worked through. They point to a gigantic problem, a massive failure to employ the machinery of intellection. A huge waste of time and thought. The Japanese have a word for such waste in manufacturing. They call it *muda*. It causes as much trouble in prose as it does in manufacturing.

If business writing presents puzzles on the one hand and pretentious flapdoodle on the other, we'll never get any business done. If we don't know who the players are or what game they are playing, we really are in the marmalade. Deciphering such prose is possible, though, and using the PM, you'll soon be doing it quickly. Isn't getting on with your business twice as fast worth the effort? In working, as in the rest of life, it's a big help to know where the action is, who's kicking who. Nobody in his or her right mind wants to write prose like that we have exemplified here. Why keep tolerating it?

In the next chapter, we move from *action* to *sentence shape*, and complete the Paramedic Method.

CHAPTER 2

Sentence Shapes

The Official Style, then, builds its sentences on the verb "to be" plus strings of prepositional phrases fore and aft; it buries its verbal action in nominative constructions with the passive voice; it often separates the natural subject from the natural verb, actor from action, by big chunks of verbal sludge; it cherishes the long windup, the slo-mo opening. Add all these attributes together and you build a sentence that has no natural shape or rhythm, no skeleton to support its meaning, no natural points of emphasis or rest. Imagine an Official Stylist in a supermarket full of prepositional phrases, with a stack of "is's" at the end of every aisle. At the checkout counter, you get a shopping bag that the writer has stuffed with words, using the generative formulas just enumerated.

This chapter considers an alternative structure, sentences with a shape that dramatizes their meaning. Sentence shape has always been important in prose but now it is vital. Digital expression has made it so. Word and image intermix and contend on the computer screen, and on its printouts. More and more, we read with images, shapes, in our eyes. Electronic writing spaces offer room for new formatting practices—

experiments in shape to focus meaning. The same pressure leans on the individual sentence. It needs the visual drama that comes, as it always has in prose style, from balance, antithesis, compression, and climactic release.

Let's start simple. On the eve of World War II, John F. Kennedy, then a senior at Harvard, wrote his undergraduate thesis on the Munich crisis. He titled it: "Appeasement at Munich (the Inevitable Result of the Slowness of Conversion of the British Democracy to Change from a Disarmament Policy to a Rearmament Policy)." Or, as we have learned to visualize such prose:

Appeasement
at Munich
(the Inevitable Result
of the Slowness
of Conversion
of the British Democracy
to Change
from a Disarmament Policy
to a Rearmament Policy)

But Jack Kennedy was no ordinary undergraduate. His Dad was the American ambassador to Britain, and rich, and so Jack got his thesis published as a book. He had some professional editorial help on the book, and especially on the title, which emerged as:

Why England Slept

Three words instead of 25 (LF 88%)—this can't be all bad. But what else has happened? The editor has swapped plodding specificity for compressed power. Metaphorical power, since "sleep" represents here "the Inevitable Result of the blah, blah, blah." It also *assumes* that the reader knows the context. It leaves the subtitle out, as Kennedy might have done by just saying "Appeasement at Munich." The Official Style *always leaves the subtitle in*. It puts detail into the sentence (usually as prepositional phrases) like a shopper filling up a shopping bag. It pays no mind to shape, rhythm, climactic power. It just stuffs in the prepositional phrases. The Official Style cherishes, at its heart, *iteration* and the making of lists.

Look at this shopping baglet:

> The yardstick for success in market competition is simple. Ultimately the vitality for a company is reckoned in terms of its profits.

OK. The PM goes to work—prepositions, "is" forms, finding the real action.

> The yardstick for success
> **in** market competition
> *is* simple.
> Ultimately the vitality
> **for** a company
> *is* reckoned
> **in** terms
> **of** its profits.

Two sentences using the same "is" + prepositional phrases formula. Let's start with the first. What is this "simple yardstick" *doing*? Where's the *action*? Hidden totally from view. What do yardsticks *do*? They *measure*. What about the natural *actor* in this sentence? Who or what is doing the measuring? The yardstick? Well, that is what yardsticks do, all right. But who is using the yardstick? Who, finally, is doing the measuring? Market competition? Vitality? ("Success" seems to = "vitality.") The company? These are simple sentences but what, we are asking, *do they really mean*? Let me suggest one easy way out:

> The yardstick for success in market competition is simple~~. Ultimately the vitality for a company is reckoned in terms of its~~ profits.

A terrific improvement in *shape*.

> The yardstick for success in market competition is simple—profits.

The sentence now possesses a natural climax that reinforces its meaning, hits the nail—profits—on the head. We can see that the second sentence simply repeats the first in other words, thereby confusing it.

But can't we use an active, transitive verb, for our central action—*measure*? How about this?

> A simple yardstick measures market success—profits.

Or, as a question:

> What yardstick measures success in the marketplace? Profits.

For either revision, a Lard Factor of about two-thirds, but more important a sentence with *shape*. No slow windup, a real action verb instead of two "is's," one prepositional phrase instead of five—*and an emphatic conclusion.* Here is how it reads to the eye:

> What yardstick measures success in the marketplace? **Profits**.

Oftentimes you can *see* a natural sentence shape trying to emerge through the lard:

> All that it really means is that more and more software will be developed faster and faster and that the software will be much more reliable and easier to maintain.

A basic shape, fighting for life:

> more and more
> faster and faster
>
> more reliable
> easier to maintain.

"More and more," "faster and faster," and "easier." You always want to *build on* patterns such as this; give the eye and ear a chance to reinforce the meaning. Revision here is dead easy.

> **All that it really means is that** more and more software will be developed faster and faster and **that the software** will be much more reliable and easier to maintain.

The slo-mo "Blah blah *is that*" opening gets snipped onto the cutting-room floor, as does "that the software," yielding:

> More and more software will be developed faster and faster, and will be more reliable and easier to maintain.

The sentence has shed a few pounds (19 words instead of 30; LF 37%) but the great improvement lies in its *shape*: nicer to look at, quicker to understand, easier to remember.

Shapelessness often makes Official Style sentences literally unreadable; read them aloud with gusto and emphasis and you feel silly. Reading prose aloud, not speed-mumbling it but using an actor's care, can tell you a lot about sentence shape. So can another simple technique. Take your sentence and write it by itself on a computer screen or sheet of paper. A little wasteful on paper, this technique costs nothing on the screen. Just open a new file and focus your sentence mid-screen. Try to sketch its architecture, to chart its lines of force. If your word processor permits it, box the selection to further frame your attention. For the previous example, we might try out various presentations:

> **More and more** software will be developed **faster and faster**, and will be **more reliable** and **easier to maintain.**

Or,

<table>
<tr><td>More and more software
faster and faster,</td><td>will be developed</td></tr>
</table>

<table>
<tr><td>and will be
and</td><td>more reliable
easier to maintain.</td></tr>
</table>

When you've gotten this far, you see that underneath the conventional linear presentation that we started with, a clear geometric shape—an X—energizes the sentence:

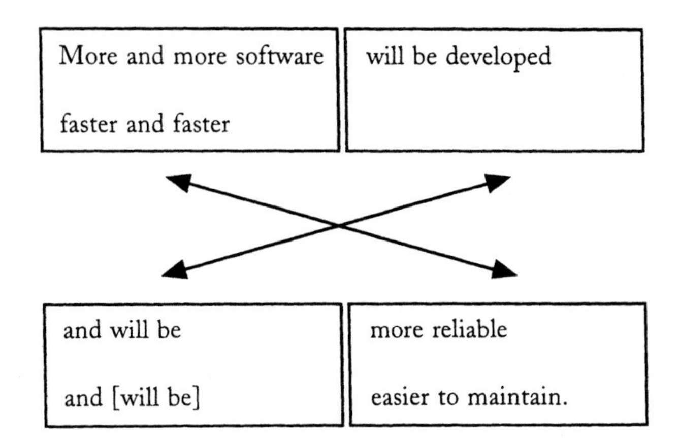

Use your draw program to draw abstract shapes that reflect its meaning: a rising line can indicate climax, a falling one the reverse; boxes of various sizes can show relative emphases; changing type font and size can help locate actor and action. We began to do this on an earlier example:

What yardstick measures success in the marketplace? **Profits**.

But we could use font manipulation to suggest relative importance of the various words. You might want the sentence to read:

What **yardstick** measures success in the **marketplace**? **Profits**.

Or:

What yardstick measures success in the ***marketplace***? **Profits**.

Use anything, however humble and unscientific, that helps you *see* how a sentence *sounds*, get a feeling for characteristic sentence shape. Let's try this method on a contrasting pair of examples.

Example 1

"They are not interested in solving the problem," said one company critic. "They are interested in keeping the bureaucracy alive, in their pensions, their perks and their power."

You can read it aloud. You can get your voice into it. It has a *shape*. But what shape?

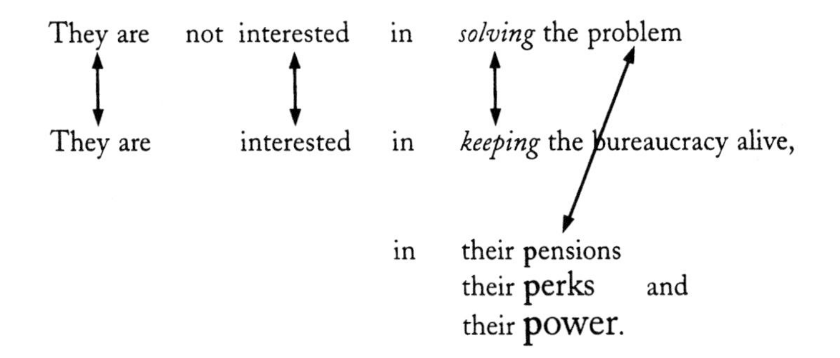

A clear pattern, stated in the negative and then repeated in the positive, with a triplet climax. And notice how the writer has used the alliteration on "p" to yoke what they are *not* interested in—the problem—and what they *are* interested in—pensions, perks, power? And how the two antithetical participles—"solving" and "keeping"—occur in the same part of each sentence, thus intensifying the contrast between them? Negative balanced by positive. And how the sentence builds upon the alliterative climax of "pensions...perks...power"?

Now this is not rocket science. It is just common sense. But how often do we heed it? More often, another shopping bag. The following paragraph opens a divisional audit within an aerospace company. *Look at it* before you read it carefully. What help does eye give to sense here?

Example 2

> We have performed a review of Transportation Control at MAD Corp. Our review was performed to determine the adequacy of and compliance with established policies and procedures, to determine if existing internal controls are sufficient to protect Company assets, and that transportation operations are administered to achieve maximum utilization of manpower, equipment and facility resources. Our review was conducted pursuant to department audit standards and included obtaining an understanding of existing procedures related to the transport function and other responsibilities of Transportation Control, completion of internal control questionnaires relating to major functions of Transportation Control, detailed tests of selected transactions, and other such tests as considered necessary.

Help from eye or ear? None at all. And yet in both the long sentences a natural shape lurks beneath the surface, waiting for a little help from its friends. Let's do it one sentence at a time, as usual—or, as an Official Stylist might put it, "pursue a stepwise methodology."

> We have performed a review of Transportation Control at MAD Corp.

The natural *action*—review—is stuffed into "performed a review." Once liberated, the rest follows easily.

> We have reviewed Transportation Control at MAD Corp.

The next sentence begins: "Our review was performed to determine...." A snap to snip the first four words and add the rest to the previous sentence.

> We have ~~performed a~~ review[*ed*] ~~of~~ Transportation Control at MAD Corp. ~~Our review was performed~~ to determine..."

Ten words instead of 16 (LF 38%) and we've *started fast*.

Onward. What are they "determining"? We learn, as soon as we've opened fast, that they are determining three things. A tripartite shape wants to break the surface.

> to determine the adequacy of and compliance with established policies and procedures,
>
> to determine if existing internal controls are sufficient to protect Company assets,
>
> and that transportation operations are administered to achieve maximum utilization of manpower, equipment and facility resources.

But the writer muffs it. You want all three parts to have the same form, so that eye and ear can register the similarities: determine X and determine Y and determine Z. Instead, the first infinitive phrase takes a simple object: "determine... adequacy." The second starts a conditional construction: "determine if...something or other." The third drops the "to determine" formula altogether and substitutes a "that" clause depending on the previous "determine"—"determine that." The tripartite shape is blurred; common sense yields to absence of mind. How focus it? Stepwise methodology again.

> to determine the adequacy of and compliance with established policies and procedures

How about setting up a "determine if" pattern? Look how it works.

> to determine if ~~the adequacy of and compliance with~~ established policies and procedures are adequate and observed

Now for Part Two. Easy:

> to determine if existing internal controls ~~are sufficient to~~ protect Company assets

And now Part Three:

> and that transportation operations are administered to achieve maximum utilization of manpower, equipment and facility resources.

Why not follow the form already established?

> to determine if manpower, equipment, and plant are used efficiently.

Now that we've lined up the three parts, *viz.*,

> **to determine if** established policies...
> **to determine if** existing internal controls...
> **to determine if** manpower...

we can envision what form the sentence should take—a *diagrammatic* one.

> We have reviewed Transportation Control at MAD Corp. to determine if:
> - established policies and procedures are adequate and observed
> - existing internal controls protect Company assets
> - manpower, equipment, and plant are used efficiently.

We've cut 55 words down to 32 (LF 42%), but much more important, we've given the information a shape the eye immediately apprehends.

Prose has been presented in conventional continuous lineation time out of mind, but the reason has also slipped out of mind. Paper was expensive—and before that, clay tablets, papyrus, stone, and parchment—and you had to use all the "white space." On a computer screen, white space is free. Diagrammatic prose is becoming much more common, part of a larger concern we are coming to call "information design."

Now for the second sentence. How do you approach, *get into* much less *digest*, a mouthful of peanut butter like this? A new PM rule to the rescue:

6. Write out each sentence on a blank screen or sheet of paper and mark off its basic rhythmic units with a "/".

Now, put the slasher to work:

> Our review **was** conducted pursuant **to** department audit standards/ and included obtaining an understanding **of** existing procedures/ related **to** the transport function/ and other responsibilities **of** Transportation Control,/ completion **of** internal control questionnaires relating **to** major functions **of** Transportation Control,/ detailed tests **of** selected transactions,/ and other such tests **as** considered necessary.

If a natural shape wants to break the surface, it has a long way to go. First, we ask how are all the parts separated by slashes related? Try a diagram that indicates the *lines of dependency* created by all those prepositional phrases and other hangers-on.

Our review **was** conducted
pursuant **to** department audit standards/
and included

obtaining an understanding **of** existing procedures/
related **to** the transport function/
and other responsibilities
of Transportation Control,/

completion
of internal control questionnaires
relating **to** major functions
of Transportation Control,/

detailed tests
of selected transactions,/
and other such tests
as considered necessary.

Aha! The skeleton, the *natural shape*, of the sentence!

The review was **conducted**...blah, blah, blah
and **included**
obtaining...blah, blah, blah
completion...blah, blah, blah
detailed tests...blah, blah, blah

The skeleton chart reveals the basic *actions* rattling around in the sentence:

conducting
pursuing
obtaining
completing
testing

Now we can begin to sort them out, step-by-step. Remember that the *actor* is *Review*. Why not put the first two—conducting and pursuing—into a single dependent clause:

Our review, which followed department audit standards,

Now what? We need, just as we did in the first sentence, a single verb to hang the three subsidiary actions on.

Our review, which followed department audit standards, **required us**

OK. What did it **require**? We'll now translate the three dependent chunks of peanut butter into plain English.

Our review, which followed department audit standards, **required us:**

- **To understand** how Transportation Control works
- **To complete** questionnaires describing its major functions
- **To test** selected transactions and other procedures.

I've cast all three requirements as infinitives ("To A, B, C") to line them up for eye and ear. And let's leave the sentence in bullet form. It is much easier to read and understand.

OK. Original and Revision:

Original

Our review was conducted pursuant to department audit standards and included obtaining an understanding of existing procedures related to the transport function and other responsibilities of Transportation Control, completion of internal control questionnaires relating to major functions of Transportation Control, detailed tests of selected transactions, and other such tests as considered necessary.

Revision

Our review, which followed department audit standards, **required us:**

- **To understand** how Transportation Control works
- **To complete** questionnaires describing its major functions
- **To test** selected transactions and other procedures.

Have I left out anything in the original, except the Official Style? Not that I can detect. And we've brought to the surface the *natural shape of the sentence*, instead of letting it languish in the depths like a chained octopus. Bottom line? Twenty-nine words instead of 52; LF 44%. But, yet again, shedding the weight is only a means to *getting the sentence into shape,* a shape easy to see and so to understand.

A Pause for Reflection

By now, we are all sweating from sweating the natural meaning and shape out of this sentence. Was this 15 minutes in a prose health club worth it? That's easy to answer. If you want to be easily and quickly understood, Yes. Especially if you want to be understood by someone who doesn't speak your special dialect of the Official Style—in this case, accountant/lawyer. Sometimes a writer wants to confuse and perplex his reader on purpose, blow smoke, as they say in Hollywood, but mostly, people doing daily business want to be understood—easily, quickly, efficiently understood.

Why would anyone write as the original example was written? They write this way because they learned to write this way—it is an accepted business practice. Just as cars used to be made in an accepted, if inefficient, way. But look at what has happened to accepted business practices across the landscape of business enterprise. They have changed. Cars are made very differently. The ram jet driving the change? Productivity. Think about the manufacturing floor after an expert in lean production has finished with it. I've tried to do the same thing in the writing space—get rid of the *muda*, as the Japanese call it. Get rid of the waste. Of course it is hard work. Learning to work differently is always the hardest kind of work. But the *muda* in the paragraph we have just finished revising drags down the writer who writes it and all the readers who read it. *Muda* in writing always amplifies itself with each new reader. It is hideously unproductive. Why, when we have learned so much about lean management in other areas, do we continue to produce, and to tolerate, it? Stay tuned.

Shaping the Sentence

The three rules which complete the Paramedic Method help us revise for a sentence shape that reflects meaning.

1. Circle the prepositions.
2. Circle the "is" forms.
3. Ask, "Where's the action?" "Who's kicking who?"
4. Put this "kicking" action in a simple active verb.
5. Start fast—no slow windups.
6. Write out each sentence on a blank screen or sheet of paper and mark off its basic rhythmic units with a "/".

7. Mark off sentence lengths in the passage with a big "/" between sentences.
8. Read the passage aloud with emphasis and feeling.

Rule 6 helps you see the shape of the individual sentence; Rule 7 helps you see the shape of a paragraph. Rule 8? It helps in all kinds of ways. If you get the giggles when you read, or start slurring your syllables, or stop and go back to the beginning, you know you're in deep *muda*. Shapeless prose has a hard time conveying strong feeling, or emphasizing its point. The emotional power in its waterlogged sentences simply seeps away. We've seen how blind the Official Style is to shape. It is deaf, too. Rule 8 turns the sound track back on.

Listen to this piece of everyday business prose:

> On the basis of the answers to these and other questions which the team might ask, I would expect the team to present us with detailed recommendations for enhancing the effectiveness of our reporting. If the recommendations are approved, we would begin to implement them immediately upon completion of the project. I would welcome a team with a broad diversity of interests, including but not limited to human resource management. Because of the focus on reporting, I would especially welcome the participation of at least one individual with a strong interest in Finance or Accounting.

Nothing serious here, no arm numbness or deep chest pain. Only flab, flab, flab, and its accompanying arrhythmia. Reading aloud emphasizes the needless repetition ("the team...the team"), the strings of jaw-breaking tongue-twisters ("recommendations for enhancing the effectiveness"), the flamenco chorus of aye, aye, ayes ("I would expect...I would welcome...I would especially welcome"). Easy fix:

Original

By answering **like these,**
~~On the basis of the answers to these~~ ~~and other~~ questions ~~which~~ the

should be able to
team ~~might ask, I would expect the team to present us with detailed~~

improvements in
recommend~~ations for enhancing the effectiveness of~~ our reporting. ~~If~~

if **be**
[T]he recommendations[,] ~~are~~ approved, ~~we~~ would ~~begin to~~

implement[*ed*] ~~them~~ immediately~~upon completion of the project~~. I ~~would~~

want **especially in**
~~welcome~~ a team with ~~a~~ broad ~~diversity of~~ interests, ~~including but not~~

and, given the
~~limited to~~ human resource management~~. Because of the focus on~~

focus,
reporting~~, I would especially welcome the participation of at least one individual with a strong interest~~ in Finance or Accounting.

Revision

By answering questions like these, the team should be able to recommend improvements in our reporting. The recommendations, if approved, would be implemented immediately. I want a team with broad interests, especially in human resource management and, given the reporting focus, in Finance or Accounting. (94 words to 45; LF 52%)

How do Original and Revision stack up in a slasher test?

Original

On the basis/
of the answers/
to these and other questions/
which the team might ask,/

Revision

By answering questions like these,/

Original

I would expect the team/
to present us/
with detailed recommendations/
for enhancing the effectiveness/
of our reporting.

Revision

the team/
should be able to recommend/
improvements in our reporting.

Original

If the recommendations are approved,/
we would begin to implement them immediately upon completion/
of the project.

Revision

The recommendations,/
if approved,/
would be implemented immediately.

Not bad. And a great improvement here not only in shape but in *voice*—crisper, clearer, emphatic, in command. More about voice in the next chapter.

How about Rule 7, checking sentence lengths?

Original

On the basis of the answers to these and other questions which the team might ask, I would expect the team to present us with detailed recommendations for enhancing the effectiveness of our reporting. (34 words)

If the recommendations are approved, we would begin to implement them immediately upon completion of the project. (17 words)

I would welcome a team with a broad diversity of interests, including but not limited to human resource management. (19 words)

Because of the focus on reporting, I would especially welcome the participation of at least one individual with a strong interest in Finance or Accounting. (25 words)

Or, in a diagram:

==================================

=================

===================

=========================

Revision

By answering questions like these, the team should be able to recommend improvements in our reporting. (16 words)

The recommendations, if approved, would be implemented immediately. (8 words)

I want a team with broad interests, especially in human resource management and, given the reporting focus, in Finance or Accounting. (21 words)

================

========

=====================

The revision improves things a little—it includes one short, emphatic, sentence—but neither passage employs a bold sentence-length strategy. There are no rules about sentence length, to be sure, except the one spice native to all human life: variety. But the principles of every human drama obtain here too: preparation, climax, release. Try an occasional climactic sentence. Try building to that climax with some longer ones.

Shaping a sentence means focusing an idea. Look at the following lost opportunity:

> We are not anxious to casually spend the company's money but our recommendation is intended to minimize the risk involved in launching a new product and a new category into an environment where there exists a vacuum of current knowledge and interest from both consumer and retailer.

Here the chick struggling out of the egg is named *Contrast*: a bad way to spend the company's money (needless research) versus a good way to spend the company's money (research that launches a product with minimum risk). The sentence as written smears this contrast over five lines. Neither eye nor ear lends the mind any help. How do we get these two powerful allies on our side? How about this:

> We don't want to waste the company's money on needless research, but informative research can save money in launching a new product, especially in an unknown market.

1. You set up an "X" pattern that brings the contrast to quick visual focus.

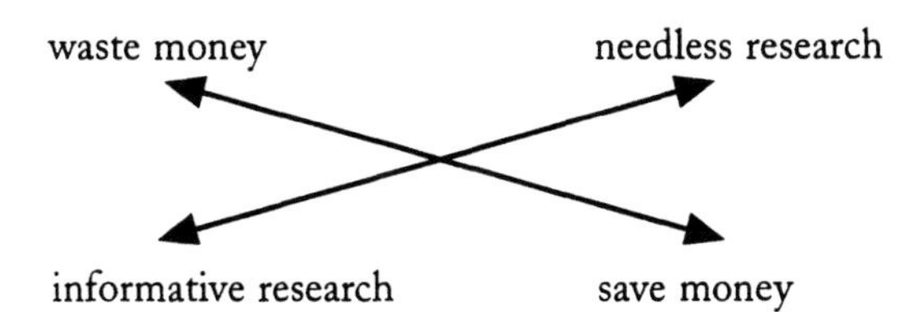

2. You put contrasted ideas in similar phrases ("waste money" and "save money," "needless research" and "informative research").
3. You invert the basic order in the second element: "money...research" becomes "research...money" to show that you have inverted the cash flow.

4. The new visual shape invites the voice to emphasize the contrast; stress "waste" and "needless" in the first element, "informative" and "save" in the second.
5. A little final guff-removal converts

> into an environment where there exists a vacuum of current knowledge and interest from both consumer and retailer

into

> especially in an unknown market.

Here are the original and the revision again:

> *Original*
>
> We are not anxious to casually spend the company's money but our recommendation is intended to minimize the risk involved in launching a new product and a new category into an environment where there exists a vacuum of current knowledge and interest from both consumer and retailer. (47)
>
> *Revision*
>
> We don't want to waste the company's money on needless research, but informative research can save money in launching a new product, especially in an unknown market. (27)

A sentence whose shape helps launch its idea, and a Lard Factor of 43%—a sentence half as long and twice as good. And think of the back pressure: writing prose that *sees* clearly may help us to *think* clearly.

THE LONG MARCH: SENTENCES-IN-LAW

The lawyers have long ruled the kingdom of long, shapeless sentences. Here is a typical one from a loan contract:

> In the event Buyer defaults on any payment, or fails to obtain or maintain the insurance required hereunder, or fails to comply with any other provision hereof, or a proceeding in bankruptcy, receivership or insolvency shall be instituted by or against Buyer or his property, or Seller deems the property in danger of misuse or confiscation, or Seller otherwise reasonably deems the indebtedness or the property insecure, Seller shall have the right to declare all amounts due or to become due hereunder to be immediately due and payable.

This sentence confuses the ordinary reader—perhaps its intent—by distending its shape. Between the opening "In the event" and the clause which completes this meaning, "Seller shall have the right...," intervene a string of qualifications.

> In the event **Buyer defaults** on any payment, or fails to obtain or maintain the insurance required hereunder, or fails to comply with any other provision hereof, or a proceeding in bankruptcy, receivership or insolvency shall be instituted by or against Buyer or his property, or Seller deems the property in danger of misuse or confiscation, or Seller otherwise reasonably deems the indebtedness or the property insecure, **Seller shall have the right** to declare all amounts due or to become due hereunder to be immediately due and payable.

We lose our way because we don't know what the qualifications apply to. The main utterance comes last instead of first, where it would orient us. Here the sentence *shape* is reinforced by the *typography*. The convention of consecutive prose obscures the shape of thought.

On an electronic screen, we can now suspend that sometimes confusing convention and reshape for thought. I won't change the wording at all, only the order and the layout:

> Seller shall have the right to declare all amounts due or to become due hereunder to be immediately due and payable, in the event:
> 1. Buyer defaults on any payment, or fails to obtain or maintain the insurance required hereunder, or
> 2. fails to comply with any other provision hereof, or
> 3. a proceeding in bankruptcy, receivership or insolvency shall be instituted by or against Buyer or his property, or
> 4. Seller deems the property in danger of misuse or confiscation, or
> 5. Seller otherwise reasonably deems the indebtedness or the property insecure.

Now let me carry the transformation one step further, revising out the formulaic repetition and enhancing the argument typographically.

> Seller can declare all amounts immediately payable, if:
> 1. **Buyer** defaults on payment or insurance
> 2. **Buyer** fails to comply with any other provision
> 3. **Buyer** goes bankrupt
> 4. **Seller** deems the indebtedness or property in danger

Whether or not the *writer* uses such a presentation, the *reader* can now recreate it as necessary. Since I'm not an attorney, I don't know how my revision would stand at law. I have simply revised for *shape*, removing the formulaic repetition and using typography to help me understand what is being written.

Happily, we have a check on my revision. I took this example from an article on plain-language laws (Michael Ferry and Richard B. Teitelman, "Plain-Language Laws: Giving the Consumer an Even Break," *Clearinghouse Review*, Vol. 14, No. 6 [Oct. 1980], pp. 522ff.). The authors give a legally sound translation, but one which I have not reviewed since I first read the article several years ago. I'll turn the page now and find out what their revision looks like:

> We can declare the full amount you owe us due any time we want to.

Ah! They have been more daring than I dared to be—but then, they are lawyers. They have a right to be. But even if we had been willing to desert the law library for the world of common sense, as they did, going through the revision we practiced would be the best way to do it. When you are revising, always proceed step by step. If you junk the sentence and substitute a paraphrase of your own—as here—though that may finally be the best thing to do, you'll not learn *why* the prose defeated you. Shape, shape, shape.

Even when they are not writing laws, or about law, lawyers cleave to the long sentence. So a typical pair from a task force report:

> In direct response to the recommendations of the 2020 Commission, the Court Technology Task Force was convened for the purpose of formulating the design, charge, and process for a permanent governing body that would oversee the planning for and implementation of technology in Anystate's trial and appellate courts.

> The task force proposed that the permanent body take the form of a standing advisory committee to the Judicial Council, to be called the Court Technology Committee, and established as its mission the promotion, coordination, and facilitation of the application of technology to the work of the Anystate courts.

OK. We'll take them one at a time, and start with PM rules 1 and 2.

> **In** direct response
> **to** the recommendations
> **of** the 2020 Commission,
> the Court Technology Task Force
> *was convened*
> **for** the purpose
> **of** formulating the design, charge, and process
> **for** a permanent governing body that would oversee
> the planning **for**
> and implementation **of** technology
> **in** Anystate's trial and appellate courts.

OK. Rule 3: Where's the action? Well, fielder's choice.

> respond
> recommend
> convene
> formulate
> govern
> plan
> implement

No shape; no clue to which action should predominate and hence who should act when we get rid of the slow windup. But the *action* clearly began with the 2020 Commission. If *we* begin there, three prepositional phrases bite the dust:

> ~~In direct response to the recommendations of~~
> The 2020 Commission recommended…

Now we have a natural subject and verbs, actor, and actions for the following clause:

> that a Court Technology Task Force be convened ~~for the purpose of formulating the~~ to design and charge ~~and process for~~ a permanent governing body ~~that would oversee the~~ to plan~~ning~~ and implement~~ation~~ technology in Anystate's trial and appellate courts.

OK. How does it look?

Original

In direct response to the recommendations of the Commission, the Court Technology Task Force was convened for the purpose of formulating the design, charge, and process for a permanent governing body that would oversee the planning for and implementation of technology in Anystate's trial and appellate courts. (47 words)

Revision

The 2020 Commission recommended that a Court Technology Task Force be convened to design and charge a permanent governing body to plan and implement technology in Anystate's trial and appellate courts. (30 words)

Cut by a third, but still pretty sludgy. Too many actors and actions. Every action described; nothing left to the reader's imagination. Back to work:

The 2020 Commission recommended that a Court Technology Task Force ~~be convened to design and~~ charge a permanent governing body ~~to plan and~~ [**with**] implement[**ing**] technology in Anystate's trial and appellate courts.

The 2020 Commission recommended that a Court Technology Task Force charge a permanent governing body with implementing technology in Anystate's trial and appellate courts. (24 words; LF 49%)

As good as I know how to make it. Now, in the second sentence, we are going to let the reader fill in a couple of gaps.

The task force proposed ~~that the permanent body take the form of~~ a standing advisory committee to the Judicial Council, ~~to be called~~ the Court Technology Committee, ~~and established as its mission the~~

Wait! Stop! Look at all those "shun" words—mission, promotion, coordination, facilitation, application. Ugh! Tongue-twisting. Easy fix.

> promotion, coordination, and facilitation of the application of technology to the work of the Anystate courts.
> = to promote and apply technology in the Anystate courts.

So, before and after for the second sentence:

> *Original*
>
> The task force proposed that the permanent body take the form of a standing advisory committee to the Judicial Council, to be called the Court Technology Committee, and established as its mission the promotion, coordination, and facilitation of the application of technology to the work of the Anystate courts. (49 words)
>
> *Revision*
>
> The task force proposed **a standing advisory committee** to the Judicial Council, the Court Technology Committee, **to promote and apply technology** in the Anystate courts. (25 words)

Lard Factor right on the money at 50%, and main actor and action stand out clearly. What drives the shapelessness in these sentences? Iterative compulsion. Every single stage in the action must be described. Nothing can be left to the reader. An understandable habit of mind in lawyers, but it destroys shape as nothing else can.

Sterling on Silver

When Warren Buffett, the most famous investor in America, was sent an Official Style revision exercise very like the ones we have been pursuing, here's how he did it.[1]

---

[1] *USA Today*, Oct. 14, 1994.

Original

Maturity and duration management decisions are made in the context of an intermediate maturity orientation. The maturity structure of the portfolio is adjusted in anticipation of cyclical interest rate changes. Such adjustments are not made in an effort to capture short-term, day-to-day movements in the market, but instead are implemented in anticipation of longer term, secular shifts in the levels of interest rates (i.e. shifts transcending and/or not inherent in the business cycle). Adjustments made to shorten portfolio maturity and duration are made to limit capital losses during periods when interest rates are expected to rise. Conversely, adjustments made to lengthen maturation for the portfolio's maturity and duration strategy lies [sic] in analysis of the U.S. and global economies, focusing on levels of real interest rates, monetary and fiscal policy actions, and cyclical indicators. (133 words)

Mr. Buffett's revision

We will try to profit by correctly predicting future interest rates. When we have no strong opinion, we will generally hold intermediate term bonds. But when we expect a major and sustained increase in rates, we will concentrate on short-term issues. And, conversely, if we expect a major shift to lower rates, we will buy long bonds. We will focus on the big picture and won't make moves based on short-term considerations. (72 words)

Mr. Buffett is obviously our kind of guy—a 46% Lard Factor and a clear statement dredged from a hopeless swamp. Not only that, but we can trust the dredger to know what the original means. Buffett on a fund prospectus is like Sterling on silver. We are free to concentrate on how he did it.

Original

Maturity and duration management decisions are made in the context of an intermediate maturity orientation. The maturity structure of the portfolio is adjusted in anticipation of cyclical interest rate changes.

Original visualized by PM rules 1 and 2

Maturity and duration management decisions
 are made
 in the context
 of an intermediate maturity orientation.
The maturity structure
 of the portfolio
 is adjusted
 in anticipation
 of cyclical interest rate changes.

OK. Standard Official Style formula:

- passive "no actor" verbs ("decisions are made")
- strings of prepositional phrases fore-and-aft
- lots of Latinate official-sounding words: maturity, duration, management, decisions, context, intermediate, orientation, etc.

What does Buffett do with it?

We will try to profit by correctly predicting future interest rates.

Lordy! He tells us who is doing the deciding: "*We* will try to profit...." That helps reduce the prepositional-phrase inventory from five to one. He makes clear the main action: "try to profit." This main *action* sticks close by the *actor*. The sentence *starts fast*. The first five of the PM rules, put into practice. And he tells us how We are going to do it: "by correctly predicting future interest rates." The Lard Factor on this segment—11 words for 30—is a stupendissimo 63%. Onward.

Original

Such adjustments are not made in an effort to capture short-term, day-to-day movements in the market, but instead are implemented in anticipation of longer term, secular shifts in the levels of interest rates (i.e. shifts transcending and/or not inherent in the business cycle).

Original as PM-diagrammed

Such adjustments
    are not made
    in an effort
    to capture short-term, day-to-day movements
    in the market,
but instead
    are implemented
    in anticipation
    of longer term, secular shifts
    in the levels
    of interest rates
        (i.e. shifts transcending and/or not inherent
            in the business cycle).

The mixture as before, but even more shapeless. How does Mr. Buffett decode it?

> When we have no strong opinion, we will generally hold intermediate term bonds.

Lordy times two—LF 70%! Again, clear *actor* and *action*: "When we have...we will...hold." Plain words—"strong opinion," "generally hold"—instead of Latinate jawbreakers. And he has complied with the Lanham Act of Prepositional Phrase Reduction by eliminating all seven prepositional phrases. Now, Act Three.

Original

Adjustments made to shorten portfolio maturity and duration are made to limit capital losses during periods when interest rates are expected to rise.

Buffett translation

But when we expect a major and sustained increase in rates, we will concentrate on short-term issues.

Now we detect a shape in Mr. Buffett's own prose. We'll align on the main *actor*.

> **We will** try to profit by correctly predicting future interest rates.
> When **we have** no strong opinion,
> **we will** generally hold intermediate term bonds.
> But when **we expect** a major and sustained increase in rates,
> **we will** concentrate on short-term issues.

A financial prospectus should tell us how a prospective management intends to act. It couldn't be better done than this. The prose takes on a clear *shape* that comes from its essential purpose. When A, we will do A+; when B, we will do B+. The rest of Buffett's revision follows this form.

> **We will** try to profit by correctly predicting future interest rates.
> When **we have** no strong opinion,
> **we will** generally hold intermediate term bonds.
> But when **we expect** a major and sustained increase in rates,
> **we will** concentrate on short-term issues.
> And, conversely,
> if **we expect** a major shift to lower rates,
> **we will** buy long bonds.
> **We will** focus on the big picture and
> **won't** make moves based on short-term considerations.

Mr. Buffett does not present his prose diagrammatically patterned, but the pattern, the *shape* of the sentences, could scarcely be clearer. **We will** or **We won't**, depending on the circumstances. He does not shrink from a term of art like "long bond" (instead of "to lengthen maturation for the portfolio's maturity and duration strategy") when he needs one, but he does not need many and he doesn't use many. This goes for most professional languages. Nor does he shrink from a common expression when it does the job—"focus on the big picture," instead of "analysis of the U.S. and global economies."

Why, it is logical to ask after this analysis, was the original not written Buffett-style to begin with? The answer comes down to *Voice* and *Authority*. The prospectus wants, above all, to sound Official and Authoritative. Mr. Buffett, with his track record, doesn't need to. To this vexed and vexing question—the voices of authority—we now turn.

CHAPTER 3

VOICES OF AUTHORITY

Nucor Corporation has been described as "the most entrepreneurial and innovative steelmaker in the world." It has led in delegating authority and flattening hierarchy as well. As its president, John Correnti, has said: "When we were a $1 billion company we had 18 people in our corporate headquarters. When we were a $1.5 billion company we had 19.... I'll fight at all costs to avoid building up a corporate hierarchy. It stifles growth, it stifles ingenuity, and it stifles that entrepreneurial spirit."[1] Here's how Correnti describes his philosophy of management:

> I can't melt steel or roll steel or sell steel or account for steel as well as those guys in that plant who do that for a living. A lot of people think that because they have the title president or executive vice president, they know more about the business than the guys on the shop floor and that's not true. I know more about the general part of it than they do, but the melter knows more about melting steel than I

[1]Quoted in *The Renaissance of American Steel*, Roger S. Ahlbrandt, Richard J. Fruehan, and Frank Giarratani (New York: Oxford University Press, 1996), p. 70.

> do, the roller knows more about rolling steel, and the sales people know more about selling steel. So you give them the encouragement to do their jobs to the best of their ability and you push it downward.[2]

Notice the *way he says it*? A believable, direct, real voice. He did not say:

> In the operation of our manufacturing facilities, it has been decided that those individuals occupying senior positions should not engage in micro management situations which result in the usurpation of melting functions from those individuals actually engaged in the process of the melting of the steel in the facilities of the Corporation. (*etc.*)

Correnti's prose gets actor and action together in the same way his management philosophy does. The shape of his description follows the shape of the business: making steel, selling it, keeping track of the sales. See the natural shape?

> I can't melt steel
> or roll steel
> or sell steel
> or account for steel
> as well as those guys in that plant who do that for a living.
>
> A lot of people think that
> because they have the title president or executive vice president,
> they know more about the business than the guys on
> the shop floor
>
> and that's not true.
>
> I know more about the general part of it than they do,
>
> but the melter knows more about melting steel than I do,
> the roller knows more about rolling steel,
> and the sales people know more about selling steel.
>
> So you give them the encouragement to do their jobs to the best of their ability and you push it downward.

---

[2]*Renaissance of American Steel*, p. 65.

Business prose like this can be spoken. *It has a voice.* The Official Style cannot speak; it can only float down from above in hierarchical layers. It is the "voice" of remote hierarchy. As such, it grates against the entrepreneurial temper of the time. Entrepreneurial prose wants to get going—no slow windups. The Official Style wants to freeze the action. Rule 8 of the PM can open your eyes quick to the philosophy of prose, and hence the philosophy of management, of a person or a company. If you can't read it aloud, with feeling, watch out.

Oddly enough, what the Official Style wants to do, above all, is to *sound authoritative.* When the flight attendant announces over the PA system that "The Captain has elected to illuminate the seat-belt sign" she wants us to say to ourselves, "Oh, Ooo, Ah, it's...The Captain!" "Elected to illuminate," instead of "turned on," tones up a cattle-car flight. People who have used the Paramedic Method to revise their prose often say, afterwards, "But I can't *just say that. It sounds too plain.*" If you aim to deceive, of course it does. Otherwise, saying what you have to say in plain speech audits your thinking, sees if its assets are really there. In August of 1940, at the height of the German invasion scare, Winston Churchill wrote a memo to his colleagues on an even more pressing matter: putting voice back into their prose.

> I ask my colleagues and their staffs to see to it that their Reports are shorter.... Let us have an end of such phrases as these: "It is also of importance to bear in the mind the following considerations..." or "Consideration should be given to the possibility of carrying into effect...." Most of these woolly phrases are mere padding, which can be left out altogether, or replaced by a single word. **Let us not shrink from using the short expressive phrase, even if it is conversational.** [emphasis mine]

All those Whitehall mandarins shrank from the short expressive phrase because it didn't sound authoritative. Nowadays, authority is not, or ought not be, so stuffy. We want to hear the straight gouge. The short expressive phrase speaks with authority as never before.

Look at the license disclaimer from the last piece of software that you bought. Then compare it to this one:

> If [SmoothFlow] doesn't work: tough. If you lose millions because of a [Smoothflow] mess up, it's you that's out the millions, not us. If you don't like this disclaimer: tough. We reserve the right to do the absolute minimum provided by law, up to and including nothing. This is basically the same disclaimer that comes with all software packages, but ours is in plain English and theirs is in legalese. We didn't want to include any disclaimer at all, but our lawyers insisted.[3]

Which disclaimer do you think speaks with more real authority?

Or remember the last corporate Annual Report you read, with its Official Style evasions. Compare it to this:

> You already know the Bad News about our past fiscal year. We were wrong about chickens. The chicken market did not recover from salmonella publicity and we entered a sharp chicken depression. We lost money in chickens—our worst in history. And the poor performance was mostly our fault. [*Wall Street Journal*, 9/30/88]

Bureaucratic language is Official Style language. Entrepreneurial language tells us that we lost money in chickens and it was our fault. The short, expressive phrase, even if conversational.

All of us want our writing to project authority. How to project it varies with time and place, obviously, but it does not come simply by falling backwards into the Official Style like a scuba diver entering the water. We've just looked at some good examples of management authority—Warren Buffett, in the last chapter, here Correnti, Churchill, and two voices of management candor. Now we'll look at some failures, starting small and funny and building up through small and sad to a Royal Proclamation.

---

[3]Quoted in *Wired*, January 1994.

The Bureaucratic Voice

Consumer Reports reports that the Isuzu Trooper flips over in tight turns. The National Highway Traffic Safety Authority found that it didn't. But what did their spokesperson say? "The question was whether there is a defect that affected its propensity to roll over, and we said no, we can't see that there is a defect here." Official Style and colloquial candor mix into explosive nonsense. The propensity to roll over is not compromised by any defect: Roll over, baby.

The sound bite, a variation of the apothegm, or short pithy saying, does not come naturally to most of us. Usually it comes out as "This is the most unthinkable alliance that became thinkable in a very short time" (of Apple getting together with IBM). People often instinctively backpedal into the Official Style. The president of an ice cream company, recruited after much ballyhoo to replace the two original founders, said when asked why he had bailed out after only 20 months on the job: "When you couple the marketplace challenges with the predictably tough demands associated with succeeding founders, the need for accelerated succession is clear." The Hippocratic Oath cautions doctors to first, Do No Harm. The sound-bite world cautions you first Not to Sound like a Nitwit.

The Job Application

When you apply for a job, you want to display your authority, your ability to do that job. Here are two examples that display disability instead. The first candidate applies for a job as editor in a publishing firm.

> I have always been fascinated by Southern California; the mild climate, the diverse people, the relaxed lifestyle, all enveloped in an atmosphere of challenge and competition. I have been involved in the same line of work for the past four and a half years. Within that short time span excelled to the highest level I could in the firm. With these two factors combining, it's now time for me to

make a career change to one that's more challenging as well as a geographic change that's more appealing. As my resume will show, I've developed certain skills that can easily be transferred from my former line of work into the publishing field.

How was the letter was read by its recipient?

I have always been fascinated by Southern California; the mild climate, the diverse people, the relaxed lifestyle, all enveloped in an atmosphere of challenge and competition.

> [Good grief! Blowing smoke at me in the first sentence.]

I have been involved in the same line of work for the past four and a half years.

> [OK. The resume says a degree in "mortuary science" and four years as an undertaker. Wordy way to say it, but I see why he wants to change jobs.]

Within that short time span excelled to the highest level I could in the firm.

> [Fatal Error! This guy not only thinks in clichés, he doesn't know the basic rules of English grammar. He is applying for a position as an *editor*???]

With these two factors combining, it's now time for me to make a career change to one that's more challenging as well as a geographic change that's more appealing.

> [Huh? Which two factors? A geographic change more appealing than what? Again, this guy can't put a simple sentence together.]

As my resume will show, I've developed certain skills that can easily be transferred from my former line of work into the publishing field.

> [Huh? What's that again? Undertaking to publishing? Embalming dead novels? Burying old authors? Running reputations into the ground?]

You can't rewrite such a letter because there is no acceptable way to say that "Four years of burying people in Cincinnati has made me want to edit books in Los Angeles." But it would be fun to try. If you could write three sentences like this together, you might even get the job.

Now a sadder case altogether. A candidate interviews for a position as securities analyst. The interview went well enough, but the writing samples sent him down in flames. First, an application letter.

> Pursuant to my phone inquiry, enclosed is a resume as recommended addressed to you. Please regard the resume and letter as a request to interview for TrustCo's analyst position.
>
> "...alas if only to place on paper the body and the soul." The following qualifications are in consideration of the analyst position. In December, my Masters of Business in Finance at California State University, Lone Pine, was completed. In addition, working full-time and starting the Financial Analyst Program allowed progress towards my three year goal to be a securities analyst.
>
> The quote is addressing the non-tangible attributes which separate the wheat from the chaff. The qualities of character, the attributes of hard and long work, high energy, quick and continued learning and a special aptitude for the securities industry are what I can offer. If other qualities are desired, please tell me so I can either convey or look to acquire them. Through reading securities industry biographies, many great achievers found focus and direction in their later twenties.

What's going on? First, though maybe not most important, there are mistakes. The first paragraph is couched in standard business formulas, but they go badly wrong. Pursuant to Rule 7 of the PM, First sentence first, and we should ponder it. The person who received the letter did precisely this, then put a question mark in the margin. "Pursuant to" Rule 7 of the PM, we'll put it on its own sheet of paper.

Pursuant to my phone inquiry, enclosed is a resume as recommended addressed to you.

Five actions take place:

pursue
inquire
enclose
recommend
address.

But the writer blurs their causal relationships, as we have seen happen so often in earlier examples. He enclosed the resume not in response to his inquiry about the job, but in response to the company's request. And we are not sure who did enclose it. "Recommended" floats without the assistance of punctuation and so the actor is blurred, too. Securities analysis is *about complex relationships*. If you can't keep a simple sentence straight, how will you do portfolio analysis? It is so easy to set straight: "As you suggested on the phone, I enclose a resume, addressed to you." Now the next sentence:

> Please regard the resume and letter as a request to interview for TrustCo's analyst position.

Another question mark in the margin. What letter? This letter? Some other letter, left out by mistake?

The writer then changes style, from Official to colloquial: "...alas, if only to place on paper the body and the soul." Huh? Three sentences, three mistakes. Now back into the Official Style:

> The following qualifications are in consideration of the analyst position.

PM Rule 3: Where's the action? Who's doing what to whom? The sentence, as written, makes no sense. He or she means: I present the following qualifications for you to consider, but again the Official Style formula blurs the action.

> In December, my Masters of Business in Finance at California State University, Lone Pine, was completed. In addition, working full-time and starting the Financial Analyst Program allowed progress towards my three year goal to be a securities analyst.

The passive voice strikes again. Who "completed" what? And how can "working full-time..." *allow progress* towards something? Again, he cannot focus actor and action in a simple sentence.

The last paragraph tries to change style again, lurches back toward an entrepreneurial directness.

> The quote is addressing the non-tangible attributes which separate the wheat from the chaff. The qualities of character, the attributes of hard and long work, high energy, quick and continued learning and a special aptitude for the securities industry are what I can offer. If other qualities are desired, please tell me so I can either convey or look to acquire them. Through reading securities industry biographies, many great achievers found focus and direction in their later twenties.

The question mark in the margin comes above the first sentence. Non-tangible attributes which separate the wheat from the chaff? The next sentence, in good Official Style, postpones the verb to the end and blurs it: "are what I can offer." And what about the last sentence? He says that many great achievers found focus by reading securities industry biographies. Is this what he means? Or rather, that by reading such biographies, he learned that many people did not become securities analysts until in their late twenties?

This paragraph, like the whole letter, shows two styles struggling for a writer's voice, two ways to project authority: the Official Style and, we might call it, the Entrepreneurial Style. The prose is trapped between two conflicting electrical fields. As a result, the writer presents just the wrong portrait of self: a person who cannot keep complex relationships straight, who cannot discriminate actor from action. A person precisely disqualified for the job.

The follow-up letter was even worse. I'll just quote a sentence or two, with the recipient's comments in brackets:

> Enthusiasm, intelligence and an yen [an *yen*?] for the investment business are arguably the most valuable indicators of a perspective [prospective] member of your team. It was encouraging to hear the strategic vision [*hear* a *vision*?] that was TrustCo's. Enclosed is a letter regarding the explanation [awful grammar and spelling] of a recent decline in the technology sector to a client. It is not often that I thoroughly [ditto] enjoyed an interview, but it is not often when I am asked to freely espouse my opinions on the economy.

A sample letter written by this candidate to a disgruntled TrustCo client was worse still. Yet the candidate was not a hopeless loser, though he came across as one. As one interviewer said, "He answered clearly all questions in our interview, and displayed as much investment knowledge and judgment as any other candidate." He was trapped in a stylistic crossfire, grabbing clichés from first one style and then the other. He had never been asked to analyze his prose as he had trained himself to analyze securities. His prose voice, just when it demands authority, ruins him.

The Royal Proclamation

Now for something completely different—a proclamation from the king of Cyrene. He, too, has investments on his mind. A Royal Proclamation deserves its own special pages.

CYRENE

ROYAL DECISION IN FAVOUR OF FOREIGN INVESTORS

In an effort to facilitate foreign investments in Cyrene His Majesty King Anon II addressed the following message to His Prime Minister

Economic development has always been and still is Our major preoccupation. It is all at once the indication of our society's cultural and intellectual level and one of the dynamic agents behind its promotion and prosperity. We have come to realize early enough that regardless of how great the efforts of the State are, Our goal cannot be fully attained without the massive contribution of the private sector whose action constitutes, particularly in the form of financial investment and know-how, one of the foundations of the development We wish for.

We have also come to realize for quite some time now that this contribution of the private sector could be effective only if it were fostered and secured of a legitimate degree of success.

With this in mind, we have taken or induced the taking of numerous measures which, in their totality, constitute Our Investment Codes.

The advantages offered by these Codes are obvious inciting factors which have not failed to produce their effects.

However, in view of the scope of the advantages she grants, Cyrene is falling quite short of her legitimate and reasonable expectations.

This inadequacy finds its major cause in the innumerable administrative procedures which, though necessary, are so slow as to discourage the most willing and best intentioned investors. Even when complete, files remain for months in the various departments while the interested parties await in total ignorance of the outcome.

Our economy can only suffer from this procedure which goes counter to our purpose.

We, therefore, have decided to put an end to that. Hence-forward, any duly constituted file consisting of an investment project shall be considered as approved by the Administration when, two months from the day of its being handed in, no action has been taken. In case the file is rejected, the administrative decision shall be duly justified.

This measure — to be implemented immediately — shall be part of the provisions of all our Investment Codes where it is to be inserted.

Meanwhile, this measure shall constitute the object of a circular issued by the Prime Minister and sent out to all the State agents. Likewise, it shall be made known to the public by all appropriate means.

ANON II
King of Cyrene

If the Official Style belongs anywhere, it belongs in a royal proclamation. This one, however, bogs down in the same thing that drags down the foreign investment it seeks to encourage: bureaucracy. Kings should speak royally, with grandeur and power. Instead, Official Style shopping bags:

> We have come to realize early enough that regardless of how great the efforts of the State are, Our goal cannot be fully attained without the massive contribution of the private sector whose action constitutes, particularly in the form of financial investment and know-how, one of the foundations of the development We wish for.

But the entrepreneurial spirit that issued the proclamation cherishes, deep down, an entrepreneurial voice as well, and it peeps out near the end.

> We, therefore, have decided to put an end to that.

Get rid of the therefore and you have a sentence brimming with real Royal Authority. You get the feeling, all the way through, that this entrepreneurial voice yearns to be liberated. Can we create a voice for an entrepreneurial king? Let's give it a try.

Bureaucrat:

> Economic development has always been and still is Our major preoccupation. It is all at once the indication of our society's cultural and intellectual level and one of the dynamic agents behind its promotion and prosperity. We have come to realize early enough that regardless of how great the efforts of the State are, Our goal cannot be fully attained without the massive contribution of the private sector whose action constitutes, particularly in the form of financial investment and know-how, one of the foundations of the development We wish for.

Entrepreneur:

We have always stressed economic development. It both indicates our society's cultural level and promotes its prosperity. We realized early that, however great the State's efforts, development required massive private financial investment and know-how.

Bureaucrat:

We have also come to realize for quite some time now that this contribution of the private sector could be effective only if it were fostered and secured of a legitimate degree of success.

With this in mind, we have taken or induced the taking of numerous measures which, in their totality, constitute Our Investment Codes.

The advantages offered by these Codes are obvious inciting factors which have not failed to produce their effects.

Entrepreneur:

We also know that private investment has to be fostered and protected. Our Investment Code has done both.

Bureaucrat:

However, in view of the scope of the advantages she grants, Cyrene is falling quite short of her legitimate and reasonable expectations.

This inadequacy finds its major cause in the innumerable administrative procedures which, though necessary, are so slow as to discourage the most willing and best intentioned investors. Even when complete, files remain for months in the various departments while the interested parties await in total ignorance of the outcome.

Our economy can only suffer from this procedure which goes counter to our purpose.

Entrepreneur:

But it hasn't fulfilled our expectations. Bureaucratic red tape has sabotaged it.

Bureaucrat:

We, therefore, have decided to put an end to that. Henceforward, any duly constituted file consisting of an investment project shall be considered as approved by the Administration when, two months from the day of its being handed in, no action has been taken. In case the file is rejected, the administrative decision shall be duly justified.

This measure — to be implemented immediately — shall be part of the provisions of all our Investment Codes where it is to be inserted.

Meanwhile, this measure shall constitute the object of a circular issued by the Prime Minister and sent out to all the State agents. Likewise, it shall be made known to the public by all appropriate means.

Entrepreneur:

We have decided to put an end to that. Effective immediately, any duly constituted investment project shall be considered as approved by the Administration when, two months from the day it is handed in, no action has been taken. This change shall be part of the Investment Code and published to all interested parties.

Now the entire Entrepreneurial Proclamation:

We have always stressed economic development. It both indicates our society's cultural level and promotes its prosperity. We realized early that, however great the State's efforts, development required massive private financial investment and know-how.

> **We also know that private investment has to be fostered and protected. Our Investment Code has done both.**
>
> **But it hasn't fulfilled our expectations. Bureaucratic red tape has sabotaged it.**
>
> **We have decided to put an end to that.**
>
> **Effective immediately, any duly constituted investment project shall be considered as approved by the Administration when, two months from the day it is handed in, no action has been taken. This change shall be part of the Investment Code and published to all interested parties.**

The original uses 362 words; our revision, 121: LF 67%. All the slow windups and wind-downs removed. More vitally, the king now speaks like a Royal Entrepreneur, not a bureaucrat: fast, tough, direct. Maybe this guy really means business. His proclamation has found its natural shape:

> **We have always stressed...**
> **We realized early...**
> **We also know...**
> **We have decided...**
>
> **Effective immediately...**

The original was trapped, like our previous example, in two conflicting stylistic fields. We've removed one, allowed the other to speak in its natural voice. The king has triumphed over his bureaucracy.

The Memo

For our last experiment in voice and authority, we must descend from monarch to memo. I've numbered the sections so we can have at it piece-by-piece.

To: Sales Managers & Representatives
From: XXX
Subject: Fax Machines In Home Offices

§1 We will be leasing from All-Fax the Fax machines without the PC/Fax Expander add-on kit based upon our experience from the trial/test we conducted at Bill Smith's home office as well as our findings about using the HP Officejet all-in-one unit. Our cost to lease these machines is about $52.00 per month for the same model L525 that you are currently using in the office. I spoke with All-Fax about procuring new Fax machines as well as using the existing machines to provide you with some more information about the logistics of how this will work.

§2 For those of you who have Fax machines and are going to take them with you to your home office, you need to know the following: Simply pack the machine up and take it home; plug it into the fax wall jack, and it will work fine as is to send and receive faxes.

The machine will retain it's original setup; However, you will have to change the name/location and phone number in the machine to reflect your new location because it will be printed on outgoing faxes. To do this, you can call the technical support toll-free number below and they will walk you through an easy procedure over the telephone in a few minutes. The Toll-Free Number is: 800-222-6666. You should keep this number handy by the Fax for future reference in case of any other problems that may occur.

§3 For those of you who do not have a Fax, we will be ordering it from headquarters. {You must already have new phone/fax lines installed into your home by the telephone company prior to setting up a machine} We must order the Fax through the purchasing by preparing a purchase requisition in advance to order a machine and get it scheduled for installation with All-Fax. I will handle all of the paperwork for you on this end, but I will need to know a few things to prepare it.

> §4 I will need to know from you: Your move-out date, the dates when you will be able to be at home for the installer to arrive and setup the machine (the machine may come in the mail in advance depending upon the arrangements for a particular location), and your new address and telephone numbers for possible contact by All-Fax for directions or to confirm their appointment. We will try to have machines in everybodys offices that need them as soon as possible. Please call me with your information as soon as you know the specifics and your schedule. Thank You!!
>
> If you have any questions, please call me at 1212.

Your office voice needs to be, in the office, as authoritative as the king of Cyrene's needs to be in his kingdom. The same vital ingredient stands threatened in proclamation and memo—administrative efficiency.

> §1 We will be leasing from All-Fax the Fax machines without the PC/Fax Expander add-on kit based upon our experience from the trial/test we conducted at Bill Smith's home office as well as our findings about using the HP Officejet all-in-one unit. Our cost to lease these machines is about $52.00 per month for the same model L525 that you are currently using in the office. I spoke with All-Fax about procuring new Fax machines as well as using the existing machines to provide you with some more information about the logistics of how this will work.

The first sentence comes out like toothpaste: no shape, no emphasis, no punctuation.

> We will be leasing from All-fax the Fax machines without the PC/Fax Expander add-on kit based upon our experience from the trial/test we conducted at Bill Smith's home office as well as our findings about using the HP Officejet all-in-one unit.

This voice cannot pause to discriminate. Let's think for a minute about how the voice wants to emphasize what is important. It should look like this:

1. We will be leasing new fax machines from All-Fax.

   [Bingo. Lead with the essential point.]

2. They will not include the PC/Fax Expander add-on kit since the trial at Bill Smith's home office, as well as our experience with the HP Officejet all-in-one unit, showed that we didn't need it.

   [Qualifying detail, but with emphatic elements at beginning and end. The voice has somewhere to go, can end with an emphatic "we didn't need it."]

3. These new machines, the same model L525 that we use now, lease for $52 per month.

   [Bottom line; again, a strong finish with the dollar amount.]

4. ~~I spoke with All-Fax about procuring new Fax machines as well as using the existing machines to provide you with some more information about the logistics of how this will work.~~

   [Needless detail—out.]

Now for Section 2:

> §2 For those of you who have Fax machines and are going to take them with you to your home office, you need to know the following: Simply pack the machine up and take it home; plug it into the fax wall jack, and it will work fine as is to send and receive faxes. (54 words)
>
> The machine will retain it's original setup; However, you will have to change the name/location and phone number in the machine to reflect your new location because it will be printed on outgoing faxes.

To do this, you can call the technical support toll-free number below and they will walk you through an easy procedure over the telephone in a few minutes. The Toll-Free Number is: 800-222-6666. You should keep this number handy by the Fax for future reference in case of any other problems that may occur.

The first paragraph makes sense; it just runs on at the mouth, telling us about every single stage in the process. Does this revision leave anything important out?

If you already have a Fax machine, simply take it to your home office and plug it into the fax wall jack. It will work fine. (26 words: LF 52%)

[I wrote for voice here. I wanted that emphatic short sentence at the end.]

Same problem in the second paragraph:

Original

The machine will retain it's original setup; However, you will have to change the name/location and phone number in the machine to reflect your new location because it will be printed on outgoing faxes. To do this, you can call the technical support toll-free number below and they will walk you through an easy procedure over the telephone in a few minutes. The Toll-Free Number is: 800-222-6666. You should keep this number handy by the Fax for future reference in case of any other problems that may occur.

Revision

The machine will retain its original setup but you will have to change the name/location/phone number so that your home office information will be printed on outgoing faxes. Call the technical support number (800-222-6666) and they'll walk you through the change. Keep this number handy for future reference.

Last two sections:

> §3 For those of you who do not have a Fax, we will be ordering it from headquarters. {You must already have new phone/fax lines installed into your home by the telephone company prior to setting up a machine} We must order the Fax through the purchasing by preparing a purchase requisition in advance to order a machine and get it scheduled for installation with All-Fax. I will handle all of the paperwork for you on this end, but I will need to know a few things to prepare it.
>
> §4 I will need to know from you: Your move-out date, the dates when you will be able to be at home for the installer to arrive and setup the machine (the machine may come in the mail in advance depending upon the arrangements for a particular location), and your new address and telephone numbers for possible contact by All-Fax for directions or to confirm their appointment. We will try to have machines in everybodys offices that need them as soon as possible. Please call me with your information as soon as you know the specifics and your schedule. Thank You!!

The writer is determined to explain every detail, important or not. Instead of disciplined thought, simply—Squeeze tube: Out comes toothpaste. I've put the vital stuff in bullets:

> If you don't have a machine, we will order one for you. To prepare the purchase requisition, and schedule installation by All-Fax, I need the following information ASAP:
>
> - Your move-out date
> - Your new address and telephone number
> - When you will be home for an installer
>
> You must, of course, install a new fax line in your home office before All-Fax can install the machine. If you have any questions, please call me at 1212.

The bullets do wonders both for the eye and for the voice. You know what to stress. The writer has thought about what he is saying, stopped just squeezing the toothpaste tube. And the Lard Factor? Original: 436 words. Revision: 207 words. Lard Factor 52%. The revision is right on the money: half as long, twice as strong. And the voice of the person handling these arrangements has some *authority*. He has thought about what he is doing and what he is saying. Controlled prose; controlled work.

A Pause for Reflection

Voice and shape vary concomitantly. Revise for one and you usually improve the other. Both model the mind of the writer. But the individual sentence models something else, too—management in miniature. The Official Style models the bureaucratic way of doing business. I've tried to suggest, through the revisions we've done up to now, an alternative entrepreneurial model for sentence management. Since entrepreneurial prose gets the job done twice as fast and usually four times as quickly, why does the Official Style still command so much high ground amongst those who manage? Why, for that matter, did so inefficient a style ever evolve to begin with? To these questions, we now turn.

CHAPTER 4

An Economics of Attention

A legendary anecdote reports that an American general once asked Winston Churchill to read the draft of a speech. "Too many passives and too many zeds," Churchill commented. Asked to explain his comment, Churchill said:

> Too many Latinate polysyllabics like "systematize, prioritize, finalize." And then the passives. What if I had said—instead of "We shall fight on the beaches"—"Hostilities will be engaged with our adversary on the coastal perimeter"?

In the previous three chapters, we have been practicing a revision method that moves from the general's Official Style 11 words to Churchill's Entrepreneurial 6. "We shall fight on the beaches" rallied England to its Finest Hour. The Official Style's "Hostilities will be engaged with our adversary on the coastal perimeter" would have elicited first a "Huh?" and then probably a "Not by me they won't." It cries out for *inaction.* By now, we've seen all the Official Style's attributes:

It hides *actor* and *action* in passive constructions.

It displaces the action from simple verb into a complex construction: "I see" becomes "A visionary ability can be obtained which permits...."

It uses a Latinate diction—all those "zed" verbs like "prioritize" and all those "shun" nouns like "prioritization."

It adores the slow sentence start, the long windup while the writer thinks up something to say: "One can easily see that in confronting a situation of this sort...."

It follows faithfully a formula of prepositional phrases + "is" + more prepositional phrases. Its "just squeeze the toothpaste tube" shapelessness gives the eye no chance to clarify the concept.

Because it offers the voice no chance to emphasize or harmonize, it is unspeakable.

And, and, and—it always takes twice as long as its Entrepreneurial translation. In the fullness of its best, it embodies the attitudes and habits of inaction, of a large, impersonal, arbitrary bureaucracy in which decisions always trickle down, mysteriously and ponderously, from the top to a horde of clerks huddled timidly under their umbrellas beneath.

A Final Pause for Reflection

Why, you might ask, does such a wasteful, unproductive prose style continue to thrive? Downsizing and lean production have gone to extreme lengths to improve productivity in other areas. Why not here? The sums you need to reckon up the savings are pretty simple: half the paper cost, half the keyboarding costs, half the time to read, three times quicker to understand—anybody can see these. And anybody can learn to revise the Official Style into a lean and dynamic Entrepreneurial one. It takes some effort, as the previous chapters teach; indeed it does. But so does learning any other new work practice, or management relationship. Compared to many other changes in the workplace, Paramedic Method revision is a snap. Yet the Official Style leaves $20 bills scattered all over the floor and no one will bother to pick them up. Why not? Many reasons contend for supremacy in an ever-shifting mixture.

Ego leads the pack. The Official Style is, or is commonly thought to be, more imposing. It speaks with organizational authority.

Fear contends with ego for the starring role. If you work in a large organization, it is fatal to stand out. You are likely to get your head chopped off, as proverbs from Persian tyrants to Mao Tse-tung caution. Don't get identified with *any* action, because if it goes wrong, you'll get the blame.

Professional mystification plays a strong role. The Official Style comes in a variety of dialects, but they are all professional languages. The language of the law came first, with the Greek and Latin patois of medical terminology a close second. Governmental bureaucracies have always cherished a special language to make them more priestly and witch-doctorish, and hence more authoritative. When the current academic and professional specializations came to the fore a century ago, they each adopted a dialect of the Official Style to prove that they were scientific. The less scientific they were, the more Official their Style. When you grow up in such a style—and every American student does—plain speech, entrepreneurial prose, makes you feel undressed. In so violently litigious a society as our own, professional mystification has come into its own as a protection against being sued.

Self-mystification plays more than a bit part. If a bureaucracy uses the Official Style for long, it begins to fool itself. After this happens—and you can see it now in all kinds of large organizations—an entrepreneurial phrase looks like *satire*. You are pointing out that the emperor has no clothes on.

Print itself must take some of the blame. It permits no direct voice. Its presentational conventions—black-and-white, continuous lineation, uniform typeface—allow only indirect emphasis. Acres of Official Style print have spawned a new kind of reading—speed reading—which encourages reading only for a key island of significance in a sea of ritualistic verbosity. Ever since Marshall McLuhan called print a "visual" medium, we've thought of it that way, but at its most profound level it doesn't work on the visual cortex. It aims to be *unnoticed* by the eye, to concentrate on the thought. A mode of presentation that actually does invoke eye and ear makes the Official Style grate on both.

Finally, the Official Style is *euphemistic*. Everyone sees this now, and laughs at it. Buzzword indexes abound. Rats become "small faunal species," smells dress up as "olfactory impacts." The U.S. Office of Education issued many years ago a self-satire which redefined commonplace words in the Official Style. "Need" became:

> A discrepancy or differential between "what is" and "what should be" (i.e. "what is required" or "what is desired"). In educational planning, "need" refers to problems rather than solutions, to the student "product" rather than to the resources for achieving that product, to the ends of education rather than to the means for attaining those ends.

But while everyone was laughing at this new kind of Official Poetic Diction, a brand new glossary of euphemisms, usually called "politically correct," prove that the habit flourishes as before. "Short" = "vertically challenged." "Fat" = "possessing an alternative body image." "Dumb" = "negatively gifted." Grammar becomes "ethnocentric white patriarchal restructuring of language." An entrepreneurial style speaks plainly about how things are. The Official Style always plays "Let's Pretend."

THE NEW SCARCE RESOURCE

Ego seems unlikely to vanish from human affairs, nor fear either, nor an unwillingness to speak plainly about ourselves. The grip of professional specialization on our lives grows ever stronger. Business organizations merge into ever larger, and therefore more bureaucratic, conglomerations. The forces generating the Official Style have all, if anything, grown stronger during the 18 years since the first edition of *Revising Business Prose* appeared. But a countervailing force has appeared during this period, too, and especially during the last decade. That force, which I call the economics of attention, supplies a completely different context and expressive space for the Official Style, and contradicts it at every point.

Different context first. If we are indeed moving from an economy based on stuff to an economy based on information, the now commonplace "information economy," an odd change has taken place. Economics, conventionally, describes how we allocate scarce commodities. But information is not a scarce commodity in our information economy. We're drowning in it. What *is* scarce is the attention we can bestow on the information drowning us. Use whatever terms you like—raw data v. massaged data, data v. information, knowledge v. wisdom—to discriminate between information in its raw form and information put into a form where we can use it productively. But these various pairs of terms point to a vital difference: information has value only when it works for us, and what makes it work for us is *human attention*. Someone has to put that information in a form we can easily understand, and we have to pay attention to the result which their attention has made comprehensible. Both the person who prepares the information and the person who consumes and acts upon it (the writer and the reader, we may for our purposes call them) expend the scarce resource of an information society—human attention. An information economy really should be called an economy of attention, and that's what we'll call it here.

In an economy of attention, a fundamental transformation takes place. In a goods economy, the central artifact is physical stuff. That stuff provides the controlling value. *Things* dominate. The language which describes those things has value only as it describes the things. The people who deal with things—dig materials from the earth's crust, process those materials, make stuff out of them, occupy center stage. The language used to describe them plays a secondary role. If you can use mathematical formulas, that is best. If you have to use words, use words that point directly to the things. But the *words* come second to the *things*. This relationship inverts in an economy of attention. Words *allocate attention*. They mediate between information and the people who use it. They now stand center stage, and stuff moves to the wings.

In a goods economy, suppose somebody writes an Official Style sentence that takes too long to tell its tale, and confuses the tale into the bargain. Well, so what? The stuff gets through finally, and that's the important thing. If you can save money *making the stuff*, good for you. That matters. But nobody saves any real money making *words* more efficient. They are a derivative function. This assumption profoundly mistakes how even a goods economy works, as our revisions in earlier chapters have shown. But the same assumption proves catastrophic in an economy of attention. When you waste words, you are wasting the principal scarce resource—attention. The Official Stylist wastes both his own attention, in adding lard, and the reader's attention as well. Attention is what, to use the popular phrase, *adds value* in an attention economy. The pompous posturing of the Official Style attacks productivity at its heart—where data is converted into productive, usable information.

We can now introduce the most compelling reason why the Official Style continues to thrive. *Because we don't think words matter*. They don't matter enough to make instruction in them a main concern. And so we train ourselves not to pay attention to them. Writing badly becomes a matter of pride; hire someone else to do it if your secretary can't. A great deal of pious cant today proclaims that writing is important, but we don't believe it. In an economy of attention you had better believe it. That's where the big gains in productivity will emerge.

Revising prose, then, becomes a central concern in an economy of attention. Prose styles constitute a microeconomics of attention. They model how we order the stream of facts and concepts bombarding us. You cannot afford not to know how sentences work. The Paramedic Method does not provide a complete science of prose style. Of course not. But it does provide *a way in*, a place to begin. It shows you *how to pay attention to a sentence*. PM revision asks that you look AT the prose surface, rather than THROUGH it. Once you know how to do that, you can begin to teach yourself through your daily experience.

Companies nowadays spend zillions of bucks putting their employees through sensitivity training, creativity-awakening sessions, group heroic endeavors of an Outward Bound sort, Zen-based therapies, and other efforts to lighten up on the one hand and buckle down on the other. Amidst all this zeal, a few dollars would be well spent in reflecting on written language as a microeconomics of attention. Prose analysis is cheaper and, one would think, easier on the nerves as well.

The information society and its economics of attention has come into being, both creating and created by a new writing space, the digital computer. Here, too, the Official Style comes under pressure. I can't illustrate the dynamic and three-dimensional expressive space of digital multimedia in a book, but I can begin to describe it, at least. Let's align its key attributes with the Official Style and see what happens.

Printed text is fixed, unchangeable. That gives it authority. To publish something is to make a final statement. Father Walter Ong, a prominent media scholar, calls print "contumacious," because you can't argue with it. It just tells you. The authority of print has a long pedigree. It comes from the

Renaissance scholars who created authoritative editions of standard classical texts. These were fixed, the last word.

Electronic text overthrows this fixity and the authority which comes with it. Electronic prose is volatile, changeable. It offers at best only a temporary printout of a continuing process of thought. The reader can argue with it, change it, rearrange it, interact with it in a variety of ways. If you interact with the Official Style, start asking it to explain itself, it rapidly comes unglued. "Huh?" "I don't follow." What do you mean by "facilitate enhanced boundary expansion conditions"? "Widen the discussion?" If you can query the writer, or ask that the point, if there is one, be gotten to a little more quickly, you've begun to prick the Official Style's gas bag.

Multimedia wise persons have been prophesying the decline of text in favor of images for some time now, but the Internet remains primarily a textual medium. And the prose styles there are characteristically colloquial, informal, low style. The Internet readily exposes the stuffing in the Official Style's shirt.

On the Net, you soon find yourself outside your own professional language, and often inside the language of some other profession. Exclusionary professional languages don't fare so well in this free-for-all as they do in the print world. They are always bouncing off one another, and off the colloquial, "entrepreneurial" style of everyday networking. If you want to preach to more than the choir, you adjust your language to suit the medium, and the Official Style doesn't suit the Net at all.

One of the clichés of media discussion argues that people don't like to read text on a screen. I've got my doubts about this, but certainly long, long, complex sentences fare especially badly there. The digital expressive space encourages you to use your eyes, notice shapes—in fact, it trains you to notice them. Shapeless prose is *seen* as such. You have to make your point quickly in a Web site. An Official Style sentence is still in slow windup when the digital reader has gone elsewhere. And prose can talk now, too. If you think a shopping bag of Official Prose sounds boring when you read it, ask your computer synthesizer to read it. One voice (most computers now offer several performing voices) sounds more ridiculous than the next. The reader of digital text will be, increasingly, trained in information design—the union of word, image, and sound. The Official Style is a design disaster. Ugly, ugly, ugly.

Now other styles compete for attention in an expressive space different from the protected precincts of print. Typography, in the digital multimedia space, has become kinetic. It has started to dance. Letters fly across the expressive stage. They come into a sentence word by word. Text has blended into choreography. I have no idea where these developments will lead, though I doubt that they can fully substitute for the consecutive text of abstract conceptual reasoning. But they certainly sensitize the eye to *shape* as a fundamental aspect of text. The more you look at the Official Style, the uglier it gets.

Bureaucracies are by nature monopolistic organizations, and they generate monopolistic prose. It doesn't have to compete with anything for attention, because the bureaucracy owns the expressive space. All this changes in the digital world. Digital multimedia create a competitive *marketplace of attention*. Word, image, sound, all play off one another and compete for our attention. Ugly, lumpen prose can't compete.

Shape becomes vital when text moves into three dimensions, as it does in the digital universe. So used are we to text as flat that we can hardly conceive it to be otherwise. Yet the multimedia textual space is one you fly into and through; you *fly over* the field of meaning. You see this flight into a new expressive space continually in the digital special effects that dominate TV commercials. Computer graphics has created a new discipline of scientific visualization which specializes in dramatizing conceptual and mathematical relationships in the three-dimensional space of ordinary behavior. Imagine a report projected into three dimensions, so that you fly farther into it for greater detail, more specificity, complexity and qualification. The ability to reformat conventional 2D text constitutes, by itself, a powerful tool for clarifying meaning and speeding up comprehension. We've seen several examples of this visual power in earlier chapters. Text in a three-dimensional field will create completely new patterns of dynamic visual apprehension. A sense of prose sound and shape will be vital. Deaf and blind styles cannot thrive in such an expressive space. In such an economy of attention, they simply cannot compete.

OK. The Official Style looks uncompetitive, whether we are talking about the macroeconomics of attention, where words and things have changed places, or the microeconomics of attention created by digital expressive space. It is a profoundly unadaptive way to communicate. And it conflicts with the current business climate in a third profound way—the philosophy of management. I read in a recent issue of the *Harvard Business Review* the advice of a French business guru that managers must do away with "the current corporate culture of command, compartmentalization, and control." As well as exhibiting a commendable taste for alliteration, he reiterates a commonplace. Hierarchical management has given way to free creative play. Hierarchy is a basic human attribute and I'm not sure we ought to banish it quite so resolutely as some gurus now recommend, but clearly a basic change in management philosophy has occurred. Decisions are being pushed down the ladder to those who know most about their consequences. People working in different parts of the enterprise are encouraged to exchange information more often and more expansively. Again, the Official Style is precisely wrong for this new management. It constitutes the voice of bureaucratic ponderosity. It models the kind of behavior the management world now seeks to avoid.

So, maybe counterforces are assembling to hustle the Official Style from the stage. If so, it won't happen quick or soon, and so we return to our beginning: the need to revise it, the need to learn how to revise it without taking six months off for a course in prose style. Such revision will help in the immediate instance and it will cultivate powers of discrimination, of judgment about shape, sound, emphasis, and stylistic level. These powers, applying across the emergent domain of information design, will constitute the central marketplace where value is added. But dangers are posed by the stylistic revision we've been demonstrating, and they form part of the story, too. It is to them, and to some final ethical reflections, that we now turn.

WHY BOTHER?

The paramedic formula does work, but it works only because the style it aims to revise is so formulaic to begin with. It is a *paramedic* method, an emergency procedure. Don't confuse it with the art of medicine, with knowing about the full range of English prose styles—how to recognize and how to write them. That larger knowledge is what English composition is all about. We are talking here about a subdivision of that broader field, about only one kind of stylistic revision. Because it is only one kind, it leaves a lot out.

Most obviously it leaves out time, place, and circumstance. It aims to be clear and brief, but often, if the social surface is to be preserved, clarity and brevity must be measured out in small doses. We seldom communicate only neutral information; we are incorrigibly interested in the emotions and human relationships that go with it.

Because the Paramedic Method ignores this aspect of writing, it can get you into trouble. Well, then, you might well ask, "Why bother?" The kind of revision we've been doing is hard work. Why do it if it's only going to get us in trouble? Why sit in your office and feel foolish trying to read a memo aloud for rhythm and shape? If the Official Style is the accepted language of our bureaucratic world, why translate it into English? Why stand up when everyone else is sitting down?

We've been discussing one reason. A lot of other people are already standing up. The economics of attention, and the digital expressive space which has emerged with it, both urge us to "bother," to understand that, in prose revision, we are adding value. Persuasive rather than autocratic management urges us along the same course. And, if these reasons seem too high-toned, we can plead simple efficiency. Here's one last example. A reader of an earlier edition of this book made the main point for me:

> "Why bother?" You omitted one of the most important reasons: Cost.
>
> Two years ago our new organization needed a Constitution and By-Laws, and a By-Laws Committee was appointed for the task. They found a sample from a similar organization, made title and other changes and produced the 12-page Constitution and By-Laws. That was not sufficient, however, and they were instructed to add more "management organization" to the Constitution. But then, of course, the By-Laws didn't agree with the expanded Constitution. One By-Laws amendment corrected one disparity, but others remained. It appeared to me that when the By-Laws were expanded to match the expanded Constitution, 19 pages of that format would be required. It was getting out of hand, and the reproduction costs could break our meager treasury.
>
> A few months ago a friend told me about *Revising Business Prose*, and suggested its principles might be applied to By-Laws as much as to straight prose. I was skeptical at first but I zeroed-in on Who's Kicking Who, eliminating useless words, nonsense phrases and needless repetition. And because we had never appointed all that added management organization, I simply eliminated them. The fever was catching, and I challenged myself to get it on four pages. To do that I used the left margin for headings and maximized print density by eliminating as much white area as possible. Our "lard factor" was perhaps as high as you've ever seen.
>
> The result is our new-look Constitution and By-Laws with easily-located subjects, quicker to read with improved comprehension, and technically more accurate.
>
> One copy now costs 20 cents versus 60 cents in the original and 95 cents if we had culminated the expansions. At our most recent meeting on April 19, 1980, it was adopted unanimously with only

one change: the part about Assessments was stricken completely because it had never been noticed before and, now that it was easy to see, the idea was unacceptable to the group!

A small instance for a large lesson. The cost—in money, time, and perplexity—exacted by the Official Style is literally and metaphorically incalculable. The reader's letter makes a second point as essential as the first—the back pressure that revision exerts on thought and imagination. Revising what we write constitutes a self-satire, a debate with ourselves. The Paramedic Method brings ideas out into the open, denies them the fulsome coloration of a special language. If the ideas are unacceptable, like the "Assessments" section of the By-Laws, we'll notice this. The Official Style encourages us to fool ourselves as well as other people, to believe in our own bureaucratic mysteries. The Paramedic Method puts our ideas back under real pressure. They can then develop and grow or—painful as this always is—find their way to the circular file. If translation into plain English reveals only banalities, it's back to the drafting board for fresh ideas. The great thing about the Paramedic Method is that it allows us to conduct this self-education in private.

We can, too, think of efficiency and writing in a slightly different, but not in the end less cost-effective, way. We live in an age of bureaucracy, of large and impersonal organizations, public and private. We're not likely to change this much. Size and impersonality seem unavoidable concomitants of the kind of global planning we'll increasingly have to do. But surely the task of language is to leaven rather than to echo this impersonality. It is a matter of efficiency as well as of humanity and aesthetic grace. We understand ideas better when they come, manifestly, from other human beings. That is simply the way human understanding evolved. It is people who act, not offices, or even officers.

The kind of translation into plain English we've been talking about can exert another kind of counterforce, as well. The Official Style, unrelievedly abstract as well as impersonal, echoes the bureaucratic preoccupation with concepts and rules. The Paramedic Method reverses the flow of this current from *concepts* back toward *objects.* It constitutes a ritual reminder to keep our feet on the ground. To the idealist philosopher's argument that the world exists only in our mind, Samuel Johnson replied by kicking a stone. The Paramedic Method does much the same kind of thing for us: Who's kicking who? The natural gravity of large organizations pulls so strongly toward concepts and abstractions that we need a formulaic counterritual. The Paramedic Method provides a start in this direction.

The language of bureaucracy, then, needs a cybernetic circuit to keep its dominant impetus toward impersonality and conceptual generalization in check. It ought to supply negative feedback, not the positive reinforcement provided by the Official Style. A counterstatement like this is more attractive, more fun—and more efficient.

The Ethical Answer

The ethical answer to "Why bother?" comes harder than the "efficiency" arguments, because it comes closely invested with questions of morality, sincerity, hypocrisy, and the presentation of self. We might begin to sketch this ethical answer by confronting the temptation head-on. Why do all of us moralize so readily about writing style? Writing is usually described in a moral vocabulary—"sincere," "open," "devious," "hypocritical"—but is this vocabulary justified? Why do so many people feel that bad writing threatens the foundations of civilization? And why, in fact, do we think "bad" the right word to use for it? Why are we seldom content to call it "inefficient" and let it go at that? Why to "clarity" and "brevity" must we always add a discussion of "sincerity" as well?

Let's start where "sincerity" starts, with the primary ground for morality, the self. We may think of the self as both a dynamic and a static entity. It is static when we think of ourselves as having central, fixed selves independent of our surroundings, an "I" we can remove from society without damage, a central self inside our head. But it becomes dynamic when we think of ourselves as actors playing social roles, a series of roles that vary with the social situations in which we find ourselves. This social self amounts to the sum of all the public roles we play. Our complex identity comes from the constant interplay of these two selves. Our final identity is usually a mixed one, few of us being completely the same in all situations or, conversely, social chameleons who change with every context. What allows the self to grow and develop is the free interplay between these two kinds of self, the central self "inside our head" and the social self "out there." If we were completely sincere we would always say exactly what we think—and cause social chaos. If we were always acting an appropriate role, we would be either professional actors or certifiably insane. Reality, for each of us, presents itself as constant oscillation between these two extremes of interior self and social role.

When we say that writing is sincere, we mean that somehow it has managed to express this complex oscillation, this complex self. It has caught the accent of a particular self, a particular mixture of the two selves. Sincerity can't point to any specific verbal configuration, to be sure, since sincerity varies as widely as people themselves. The sincere writer has not said precisely what she felt in the first words that occur to her. That might produce a revolutionary tirade or "like, you know" conversational babble. Nor has a sincere writer simply borrowed a fixed language, as when a bureaucrat writes in the Official Style. The "sincere" writer has managed to create a style which, like the social self, can become part of society, can work harmoniously in society and, at the same time, like the central self, can represent her unique selfhood. She holds her two selves in balance; from that balance emerges the "authority" we discussed in Chapter 3.

The act of writing involves for the writer an integration of her self, a deliberate act of balancing its two component parts. It represents an act of socialization, and it is by repeated acts of socialization that we become sociable beings, that we grow up. Thus, the act of writing models the presentation of self in society, constitutes a rehearsal for social reality. It is not simply a question of a preexistent self making its message known to a preexistent society. From the ethical point of view, it is not, initially, a question of message at all. Writing is a way to clarify, strengthen, and energize the self, to render individuality rich, full, and social. This does not mean writing that flows, as Terry Southern immortally put it, "right out of the old guts onto the goddamn paper." Just the opposite. Only by taking the position of the reader toward one's own prose, putting a reader's pressure on it, can the self be made to grow. Writing can, through this pressure, enhance and expand the self, allow it to try out new possibilities, tentative selves. We return here to the back pressure revision exerts. It stimulates not only the mind but the whole personality. We are offering not simply an idea but our personality as context for that idea. And just as revision makes our ideas grow and develop, it encourages us to remember the different ways we can act in society, the alternative paths to socialize the self.

The moral ingredient in writing, then, works first not on the morality of the message but on the nature of the sender, on the complexity of the self. "Why bother?" To invigorate and enrich your selfhood; to increase, in the most literal sense, your self-consciousness. Writing, properly pursued, does not make you better. It makes you more alive, more coherent, more in control. A mind thinking, not a mind asleep. It aims, that is, not to denature the human relationship that writing sets up, but to enhance and enrich it. It is not trying to squeeze out the expression of personality but to make such expression possible, not trying to obscure all records of a particular occasion and its human relationships but to make them maximally clear. Again, this is why we worry so much about bad writing. It signifies incoherent people, failed social relationships. This worry makes sense only if we feel that writing, ideally, should express human relationships and feelings, not abolish them.

From the ethical point of view, then, we revise the Official Style when it fails to socialize the self and hence to enrich it, to discipline the ego to the surrounding egos that give it meaning. This, unhappily, is most of the time. Pure candor can be soundly destructive but so can pure formula, endless cliché. When formula takes over, self and society depart. The joy goes out of the prose. It's no fun to write. And when this happens, you get those social gaffes, those trodden toes, those "failures of communication" that so often interfere with the world's business. The human feeling that has been pushed out the front door sneaks in the back. So when you cease to feel good about what you write, when you cease to add something of yourself to it, watch out!

ETHICS *AND* EFFICIENCY

When we put these two answers to "Why bother?" together—ethics and efficiency—we discover a paradoxical convergence. Cases do exist where one answer will do by itself—in the By-Laws case, for example, the "efficiency" argument is all we need—but more often than not the two kinds of justification support one another. The "efficiency" argument, pressed hard enough, comes to overlap the ethical argument and vice versa. We may, in this area of overlap, have come across the richness we feel when we use all the customary value-laden terms to describe a piece of prose—"sincere," "honest," "fresh," "straightforward," and so on. We feel that somehow ego and efficiency have come to collaborate in establishing a clarity that makes understanding a pleasure and a shared one.

At this point, the paramedic analogy breaks down. Beyond paramedicine lies medicine; beyond the specific analysis of a specific style—what we have been doing here—lies the study of style in general and its relation to human motive and behavior. Verbal style can no more be fully explained by a set of rules, stylistic or moral, than can any other kind of human behavior. Intuition, trained intuition, figures as strongly in the one as in the other. You must learn how to see.

You'll then be able to answer—situation by situation, one instance at a time, as business decisions are always reached—the fundamental question that this chapter, and this book, can only introduce. How to revise the Official Style is easy, once you know how. As we've seen, anyone can do it. The questions that generate no rules, the questions that try our judgment—and our goodness—are *When*? and *How*? If, and as, American business keeps merging into bigger and bigger firms, that is more hierarchical and bureaucratic ones, these questions will loom ever larger, and applying the rules will get ever harder. I wish there were an easier answer—a rule for when to apply the rules. But no one has ever found such a touchstone.

When a famous violinist was stopped on the street and asked the way to Carnegie Hall, he replied, "Practice!" Practice is what we have been doing together in *Revising Business Prose*. And practice shows us the only way from the Official Style to a discourse which does business in a more efficient and humane way.

APPENDIX

TERMS

You can see things you don't know the names for, but in prose style, as in everything else, it is easier to see what you know how to describe. The psychological ease that comes from calling things by their proper names has not often been thought a useful goal by modern pedagogy. As a result, inexperienced writers often find themselves reduced to talking about "smoothness," "flow," and other meaningless generalities when they are confronted by a text. And so here are some basic terms.

PARTS OF SPEECH

In traditional English grammar, there are eight parts of speech: verbs, nouns, pronouns, adjectives, adverbs, prepositions, conjunctions, interjections. *Grammar*, in its most general sense, refers to all the rules that govern how meaningful statements can be made in any language. *Syntax* refers to sentence structure, to word order. *Diction* means simply word choice. *Usage* means linguistic custom.

Verbs

1. Verbs have two voices, active and passive.
   An *active verb* indicates the subject acting:
   Jack *kicks* Bill.
   A *passive verb* indicates the subject acted upon:
   Bill *is kicked by* Jim.
2. Verbs come in three moods: indicative, subjunctive, and imperative.
   A verb in the *indicative mood* says that something is a fact. If it asks a question, it is a question about a fact:
   Jim *kicks* Bill. *Has* Bill *kicked* Jim yet?
   A verb in the *subjunctive mood* says that something is a wish, hypothetical, or contrary to fact, rather than a fact:
   If Jim *were* clever, he *would* kick Bill.
   A verb in the *imperative mood* issues a command:
   Jim, *kick* Bill.
3. A verb can be either transitive or intransitive.
   A *transitive verb* takes a direct object:
   Jim *kicks* Bill.
   An *intransitive verb* does not take a direct object. It represents action without a specific goal:
   Lori *runs* every day.
   The verb "to be" ("is," "was," and so on) is often a *linking* verb because it links subject and predicate without expressing a specific action:
   Elaine *is* a movie mogul.
4. English verbs have six tenses: present, past, present perfect, past perfect, future, and future perfect.
   Present: Jim *kicks* Bill.
   Past: Jim *kicked* Bill.
   Present perfect: Jim *has kicked* Bill.
   Past perfect: Jim *had kicked* Bill.
   Future: Jim *will kick* Bill.
   Future perfect: Jim *will have kicked* Bill.

The present perfect, past perfect, and future perfect are called compound tenses. Each tense can have a progressive form. (e.g., present progressive: Jim *is kicking* Bill.)

5. Verbs in English have three so-called infinite forms: *infinitive*, *participle*, and *gerund*. These verb forms often function as adjectives or nouns.

   Infinitive:

   To assist Elaine isn't easy.

   (When a word separates the "to" in an infinitive from its complementary form, as in "to directly stimulate" instead of "to stimulate," the infinitive is said to be a split infinitive.)

   Participles and gerunds have the same form; when the form is used as an adjective, it is called a *participle*; when used as a noun, a *gerund*.

   Participles:

   Present participle:

   Elaine was in an *arguing* mood.

   Past participle:

   Lori's presentation was very well *argued*.

   Gerund:

   Arguing with Elaine is no fun.

Verbs that take "it" or "there" as subjects are said to be in an *impersonal construction*: "It has been decided to fire him" or "There has been a personnel readjustment."

Nouns

A noun names something or somebody. A proper noun names a particular being or place—Elaine, Pittsburgh.

1. *Number*. The singular number refers to one ("a cat"), plural to more than one ("five cats").
2. *Collective nouns*. Groups may be thought of as a single unit, as in "the army," and thus take a singular verb.

Pronouns

A pronoun is a word used instead of a noun. There are different kinds:

1. *Personal pronouns*: I, me, him,...
2. *Intensive pronouns*: myself, yourself,...
3. *Relative pronouns*: who, which, that. These must have antecedents, words they refer back to. "Lori has a talent (antecedent) that (relative pronoun) Elaine does not possess."
4. *Indefinite pronouns*: somebody, anybody, anything
5. *Interrogative pronouns*: who?, what?

Possessives

Singular: A *worker's* hat. Plural: The *workers'* hats. ("It's," however, equals "it is." **The possessive is "its"—no apostrophe!**)

Adjectives

An *adjective* modifies a noun: "Lori was a *good* hiker."

Adverbs

An *adverb* modifies a verb: "Lori hiked *swiftly* up the trail."

Prepositions

A *preposition* connects a noun or pronoun with a verb, an adjective, or another pronoun: "I ran into her arms" or "The girl with the blue scarf."

Conjunctions

Conjunctions join sentences or parts of them. There are two kinds, coordinating and subordinating.

1. *Coordinating conjunctions*—and, but, or—connect statements of equal status: "Bill ran *and* Jim fell" or "I got up *but* soon fell down."
2. *Subordinating conjunctions*—that, when, because—connect a main clause with a subordinate one: "I thought *that* they had left."

Interjections

A sudden outcry: "Wow!" or "Ouch!"

SENTENCES

Every sentence must have both a subject and verb, stated or implied: "Elaine (subject) directs (verb)."

Three Kinds

1. A *declarative sentence* states a fact: "Elaine directs films."
2. An *interrogative sentence* asks a question: "Does Elaine direct films?"
3. An *exclamatory sentence* registers an exclamation: "Does she ever!"

Three Basic Structures

1. A simple sentence makes one self-standing assertion, i.e., has one main clause: "Elaine directs films."
2. A compound sentence makes two or more self-standing assertions, i.e., has two main clauses: "Elaine directs films and Lori is a tax lawyer" or "Jim kicks Bill and Bill feels it and Bill kicks Jim back."
3. A complex sentence makes one self-standing assertion and one or more dependent assertions in the form of subordinate clauses dependent on the main clause:

> "Elaine, who has just finished directing *Jim Kicks Bill*, must now consult Lori about her tax problems before she can start blocking out *Being Kicked: The Sequel.*"

In *compound sentences*, the clauses are connected by *coordinating conjunctions*, in *complex sentences* by *subordinating conjunctions.*

Restrictive and Nonrestrictive Relative Clauses

A *restrictive clause* modifies directly, and so restricts the meaning of the antecedent it refers back to: "This is the tire *that blew out on the freeway.*" One specific tire is referred to. Such a clause is not set off by commas, because it is needed to complete the meaning of the statement about its antecedent: "This is the tire"—what tire?

A *nonrestrictive clause*, though still a dependent clause, does not directly modify its antecedent and is set off by commas: "These tires, *which are quite expensive,* never blow out on the freeway." A nonrestrictive clause can be removed without changing the sense of the main clause: "These tires never blow out on the freeway."

Appositives

An *appositive* is an amplifying word or phrase placed next to the term it refers to and set off by commas: "Henry VIII, *a glutton for punishment,* rode out hunting even when sick and in pain."

Basic Sentence Patterns

What words do you use to describe the basic syntactic patterns in a sentence? In addition to the basic types—declarative, interrogative, and exclamatory—and the basic forms of simple, compound, and complex, other terms sometimes come in handy.

Parataxis and Hypotaxis

Parataxis: Phrases or clauses arranged independently, in a coordinate construction, and often without connectives, e.g., "I came, I saw, I conquered."

Hypotaxis: Phrases or clauses arranged in a dependent subordinate relationship, e.g., "I came, and after I came and looked around a bit, I decided, well, why not, and so conquered."

The adjectival forms are *paratactic* and *hypotactic*, e.g., "Hemingway favors a paratactic syntax while Faulkner prefers a hypotactic one."

Asyndeton and Polysyndeton

Asyndeton: Connectives are omitted between words, phrases, or clauses, e.g., "I've been stressed, destressed, beat down, beat up, held down, held up, conditioned, reconditioned."

Polysyndeton: Connectives are always supplied between words and phrases, or clauses, as when Milton talks about Satan pursuing his way, "And swims, or sinks, or wades, or creeps, or flies."

The adjectives are *asyndetic* and *polysyndetic*.

Periodic Sentence

A periodic sentence is a long sentence with a number of elements, usually balanced or antithetical, standing in a clear syntactical relationship to each other. Usually it suspends the conclusion of the sense until the end of the sentence, and so is sometimes said to use a *suspended syntax*. A periodic sentence shows us a pattern of thought that has been fully worked out, whose power relationships of subordination have been carefully determined, and whose timing has been climactically ordered. In a periodic sentence, the mind has finished working on the thought, left it fully formed.

There is no equally satisfactory antithetical term for the opposite kind of sentence, a sentence whose elements are loosely related to one another, follow in no particularly antithetical climactic order, and do not suspend its grammatical completion until the close. Such a style is often called a *running style* or a *loose style*, but the terms remain pretty vague. The loose style, we can say, often reflects a mind *in the process of thinking* rather than, as in the periodic sentence, having already completely ordered its thinking. A sentence so loose as to verge on incoherence, grammatical or syntactical, is often called a *run-on sentence.*

Isocolon

The Greek word *isocolon* means, literally, syntactic units of equal length, and it is used in English to describe the repetition of phrases of equal length and corresponding structure. So Winston Churchill on the life of a politician: "He is asked to stand, he wants to sit, and he is expected to lie."

Chiasmus

Chiasmus is the basic pattern of antithetical inversion, the AB:BA pattern. President John F. Kennedy used it in his inaugural address:

A	B
Ask not *what your country*	*can do for you*, but
B	**A**
what you can do	*for your country.*

Anaphora

When you begin a series of phrases, clauses, or sentences with the same word or phrase, you are using anaphora. So Shakespeare's Henry V to some henchpersons who have betrayed him:

Show men dutiful?
Why, so didst thou. Seem they grave and learned?
Why, so didst thou. Come they of noble family?
Why, so didst thou. Seem they religious?
Why, so didst thou.

(*Henry V,* 2.2)

Tautology

Repetition of the same idea in different words. In many ways, the Official Style is founded on this pattern. Here's a neat example from Shakespeare:

Lepidus. What manner o'thing is your crocodile?
Antony. It is shap'd, sir, like itself, and it is as broad as it has breadth. It is just so high as it is, and moves with its own organs. It lives by that which nourisheth it, and the elements once out of it, it transmigrates.
Lepidus. What colour is it of?
Antony. Of its own colour too.
Lepidus. 'Tis a strange serpent.
Antony. 'Tis so. And the tears of it are wet.

(*Antony and Cleopatra,* 2.7)

Noun Style and Verb Style

Every sentence must have a noun and a verb, but one can be emphasized, sometimes almost to the exclusion of the other. The Official Style—strings of prepositional phrases + "is"—exemplifies a noun style *par excellence.* Here are three examples, the first of a noun style, the second of a verb style, and the third of a balanced noun-verb mixture.

Noun Style

There is in turn a two-fold structure of this "binding-in." In the first place, by virtue of internalization of the standard, conformity with it tends to be of personal, expressive and/or instrumental significance to ego. In the second place, the structuring of the reactions of alter to ego's action as sanctions is a function of his conformity with the standard. Therefore conformity as a direct mode of the fulfillment of his own need-dispositions tends to coincide with the conformity as a condition of eliciting the favorable and avoiding the unfavorable reactions of others.

(Talcott Parsons, *The Social System* [Glencoe, Ill.: Free Press, 1951], p. 38)

Verb Style

Patrols, sweeps, missions, search and destroy. It continued every day as if part of sunlight itself. I went to the colonel's briefings every day. He explained how effectively we were keeping the enemy off balance, not allowing them to move in, set up mortar sites, and gather for attack. He didn't seem to hate them. They were to him like pests or insects that had to be kept away. It seemed that one important purpose of patrols was just for them to take place, to happen, to exist; there had to be patrols. It gave the men something to do. Find the enemy, make contact, kill, be killed, and return. Trap, block, hold. In the first five days, I lost six corpsmen—two killed, four wounded.

(John A. Parrish, *12, 20 & 5: A Doctor's Year in Vietnam* [Baltimore: Penguin Books, 1973], p. 235)

Mixed Noun-Verb Style

We know both too much and too little about Louis XIV ever to succeed in capturing the whole man. In externals, in the mere business of eating, drinking, and dressing, in the outward routine of what he loved to call the *métier du roi*, no historical character, not even Johnson or Pepys, is better known to us; we can even, with the aid of his own writings, penetrate a little of the majestic façade which is Le Grand Roi. But when we have done so, we see as in a glass darkly. Hence the extraordinary number and variety of judgments which have been passed upon him; to one school, he is incomparably the ablest ruler in modern European history; to another, a mediocre blunderer, pompous, led by the nose by a succession of generals and civil servants; whilst to a third, he is no great king, but still the finest actor of royalty the world has ever seen.

(W. H. Lewis, *The Splendid Century: Life in the France of Louis XIV* [New York: Anchor Books, 1953], p. 1)

PATTERNS OF RHYTHM AND SOUND

Meter

The terms used for scanning (marking the meter of) poetry sometimes prove useful for prose as well.

iamb: unstressed syllable followed by a stressed one, e.g., in vólve.
trochee: opposite of iamb, e.g., úse ful.
anapest: two unstressed syllables and one stressed syllable, e.g., per son nél.
dactyl: opposite of anapest, one stressed syllable followed by two unstressed ones, e.g., óp er ate.

These patterns form *feet*. If a line contains two feet, it is a *dimeter*; three, a *trimeter*; four, a *tetrameter*; five, a *pentameter*; six, a *hexameter*. The adjectival forms are *iambic*, *trochaic*, *anapestic*, and *dactylic*.

Sound Resemblances

Alliteration: This originally meant the repetition of initial consonant sounds but came to mean repetition of consonant sounds wherever they occurred, and now is often used to indicate vowel sound repetition as well. You can use it as a general term for this kind of sound play: "Peter Piper picked a peck of pickled peppers"; "Bill will always swill his fill."

Homoioteleuton: This jawbreaker refers, in Latin, to words with similar endings, usually case-endings. You can use it to describe, for example, the "shun" words—"function," "organization," "facilitation"—and the sound clashes they cause.

For further explanation of the basic terms of grammar, see George O. Curme's *English Grammar* in the Barnes & Noble College Outline Series. For a fuller discussion of rhetorical terms like *chiasmus* and *asyndeton*, see Richard A. Lanham's *A Handlist of Rhetorical Terms* (second edition, Berkeley and Los Angeles: University of California Press, 1991). For a fuller discussion of prose style, see Richard A. Lanham's *Analyzing Prose* (New York: Scribner's, 1983).

INDEX